Creating Technology Strategies

Creating Technology Strategies

How to Build Competitive Biomedical R&D

Alice M. Sapienza, DBA

Professor, Graduate School for Health Studies
Simmons College

WILEY-LISS

A JOHN WILEY & SONS, INC., PUBLICATION
New York • Chichester • Weinheim • Brisbane • Singapore • Toronto

Address All Inquiries to the Publisher
Wiley-Liss, Inc., 605 Third Avenue, New York, NY 10158-0012

Printed in the United States of America

Library of Congress Cataloging in Publication Data:

Sapienza, Alice M.
Creating technology strategies : how to build competitive biomedical R&D / Alice M. Sapienza.
p. cm.
Includes index.
ISBN 0-471-15370-2 (cloth : alk. paper)
1. Medicine—Research—Planning. 2. Biology—Research—Planning. 3. Medical instruments and apparatus industry. 4. Biotechnology industries. 5. Pharmaceutical industry. I. Title.
R852.S27 1997
610'.72—dc21 97-28682

The text of this book is printed on acid-free paper.

10 9 8 7 6 5 4 3 2 1

To Bill Curran (1925–1996),
with gratitude for starting me on this trajectory

CONTENTS

PREFACE

Why a book on strategy, designed particularly for the biomedical industry? This industry—a very-high-technology, knowledge industry—has a number of unique attributes that warrant focused attention. For instance, the life sciences and related technologies are advancing with tremendous speed; because of this, biomedical product and process life cycles are being compressed, sometimes by years; and, at least in the biopharmaceuticals sector, the competitive landscape is being radically altered by mergers and a web of alliances.[1]

As if these attributes were not sufficiently problematic, the biomedical industry also operates in a unique environment, that of health care. There may be no other situation in which:

- products and services are viewed as inalienable "rights" of the consumer;
- the consumer is actually three entities—medical professional, patient, and insurer—who may disagree among themselves;

- marvelous medical innovations stemming from government-funded health research also face tight cost constraints from the same governments;[2]
- the emotional content of so many decisions is high.

I know a number of these issues from my experience as a general manager in a Harvard University teaching hospital. And, for more than a decade, I have been consulting to R&D scientists in pharmaceutical, diagnostic, and biotechnology firms worldwide. I appreciate the struggle of preparing these companies to survive and flourish in the twenty-first century.

But, more importantly, I have discovered from my consulting that this struggle is too often exacerbated by a *gap* between scientists' understanding of corporate strategy and its implications for R&D, and the understanding by corporate nonscientists of the impact of R&D decisions on the achievement of business objectives. When there is such a gap, the necessary coherence between corporate strategy and technology strategy—the blueprint that must guide the day-to-day allocation of science and technology resources—is missing. Because of the unique attributes of the biomedical industry and the unique context of the health care environment, if there is any inconsistency between business and technology decisions, the biomedical firm is imperiled.

These are my reasons for writing a book on strategy, designed particularly for the biomedical industry. My objective is to help *change the way scientists and nonscientists think* about strategy, environment, and technology, so that the necessary consistency among corporate, R&D, program, and project decisions will be achieved. Readers who should find this material useful include, of course, those who work in the biomedical industry: (1) *members of the R&D leadership team,* who must be able to understand, design, and implement a technology strategy that supports the achievement of business objectives; (2) *corporate management,* who must better understand the role of science and technology in the attainment of strategic goals; and (3) *R&D scientists,* who must be able to assess and manage programs and projects in ways that are consistent with corporate and technology strategies.

In addition, this book should be useful to scientists and corporate managers in other very-high-technology industries, to academics charged with preparing scientific and technical professionals to manage very-high-technology firms, and to anyone concerned about the coherence between business and R&D strategies.

This book addresses a number of important needs. It explains the special context of biomedical firms and some of the pressures on the leadership of these firms to invest and manage R&D resources astutely. It provides a conceptual foundation for understanding the technology life cycle, so that science and technology resources can be appropriately invested. And it supplies tools by which the leadership team can:

- analyze the external environment, assess levels of turbulence, and make use of weak signals to preempt the competition;
- codify the competitive position of the firm, from the perspective of biomedical science and technology;
- determine the consistency of R&D investments with the firm's strategic vision;
- determine the consequences of and requirements for changing competitive position, from the perspective of biomedical science and technology;
- assess the fit of external knowledge sources (new recruits, acquired firms, etc.) with the biomedical firm;

- implement competitive technology strategy at the program and project levels.

The U.S. National Critical Technologies Panel stated: "In an environment of intensifying global competition, deployment of technology is becoming the strategic battlefield of the international marketplace."[3] This book will help to prepare business and technical leaders for such a battlefield.

NOTES

1. In 1994, the Ernst & Young annual report on the biotechnology industry noted that nearly all pharmaceutical and biotechnology companies were "forming and managing multiple relationships, . . . [that reinforce the] paradigm shift toward virtual integration" of the biopharmaceutical industry (K. B. Lee and G. S. Burrill, *Biotech 95: Reform, Restructure, Renewal,* Ernst & Young, Palo Alto, CA, p. 220).
2. I described the "irreconcilable differences" facing, especially, biotechnology medicines (A. M. Sapienza, "Irreconcilable Differences? Strategic Innovations for a Publicly Insured Market," *Technovation 12,* 5, 1992).
3. This statement comes from the National Critical Technologies Panel report to then President Bush, Arlington, VA (March 1991), p. 1.

ACKNOWLEDGMENTS

This book would never have been possible without four key people who read and reread and read again its many drafts.

For two people—Ivan Jensen, MD, and Diana Stork, PhD—this labor of love was performed for me a second time, because they also read (and reread) the drafts of my earlier book, *Managing Scientists* (Wiley, 1995). In addition to reading and commenting on the manuscript, Ivan telephoned from Denmark (usually very early in the morning) with much-needed "crits."

The third key person in the evolution of this book is Joseph Lombardino, PhD. He was in at the beginning, commenting on the prospectus. He continued to read and comment on all the drafts (sometimes on the train during his commute to and from work), leaving extremely important marginalia of encouragement.

The fourth key person is Prof. Dr. Hans-Harald Sedlacek, who spent much time trying to help me turn my illustrations into better graphics.

I am grateful to these colleagues because of their insightful questions, their invaluable comments on my material, and their most necessary support.

I also want to thank Eugene R. Langevin, MD, Chief of Radiology at St. Elizabeth's Hospital, Boston. He was most gracious in helping me better understand diagnostic imaging devices.

Finally, I appreciate the continued encouragement and assistance of Colette Bean at John Wiley & Sons. She and I began our association on this book pleasantly, in a coffee shop on Third Avenue in New York. We continue to this day in a warm and mutually respectful relationship that I value.

Creating Technology Strategies

CHAPTER ONE

ABOUT THIS BOOK

This is not intended to be a book about strategic planning for firms in the biomedical industry. Rather, it is intended to be a book that provokes a novel way of thinking about strategy from the perspective of biomedical science and technology. The purpose of this chapter is to orient the reader to the difference. The chapter describes the impetus for the book, reviews the objective and organizing model following from that objective, and provides a brief guide to the remaining contents.

IMPETUS

Few would disagree with the need to ensure a strong link between corporate strategy and research and development (R&D) activities within the pharmaceutical, diagnostic, device, and biotechnology firms that constitute the biomedical industry.[1] But I have found that there is too often a gap (Figure 1) between:

- scientists' understanding of corporate strategy and its implications for R&D, and
- the understanding by nonscientists of the impact of R&D decisions on the achievement (or nonachievement) of business objectives.

One reason for this gap is that people at corporate, R&D, program, and project levels in the organization may have incomplete or even contradictory perspectives of the role of science and technology in the attainment of business objec-

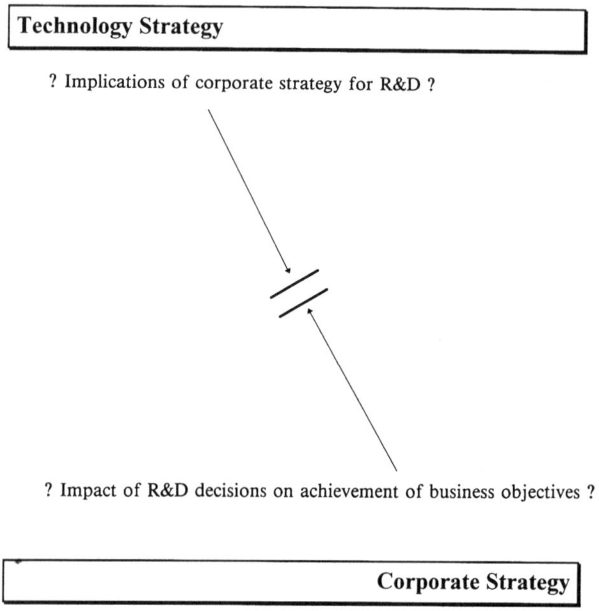

Figure 1. The gap.

tives. Another reason is that communication in the firm may be primarily top-down or be impeded because people speak different "languages" (e.g., "business" versus "science") and are rewarded for achieving different goals. Whatever the cause, the result is the same: internal consistency among corporate, R&D, program, and project decisions is missing.

Some years ago—during the so-called glory days of the biomedical industry—most products that were approved generated revenues for the company. A few products even generated enormous revenues and profits. Then, coherence (consistency) between corporate strategy and R&D activities was a less-salient issue; in fact, "strategy" was rarely mentioned. Now and for the foreseeable future, however, companies in the biomedical industry must have a defensible corporate strategy and, in that context, must allocate the right amount of R&D resources to the right people (i.e., programs and projects) at the right time.

I believe that the degrees of strategic freedom in the biomedical industry have been reduced by (Figure 2):

- cost-constraining pressures from strong buyers of biomedical products and services, especially governments and managed-care organizations, coupled with demand that new products and services demonstrate clear advantages over less expensive therapies
- compression of product and process life cycles by rapidly advancing science frontiers
- intense competition and consolidation of firms (several companies are likely to be competing for a particular product class; thus, astute research decisions and speed of development are even more crucial).

Because of these dynamics (discussed also in Chapter Three) any gap, any inconsistency or lack of coherence between business and technology decisions is truly perilous.

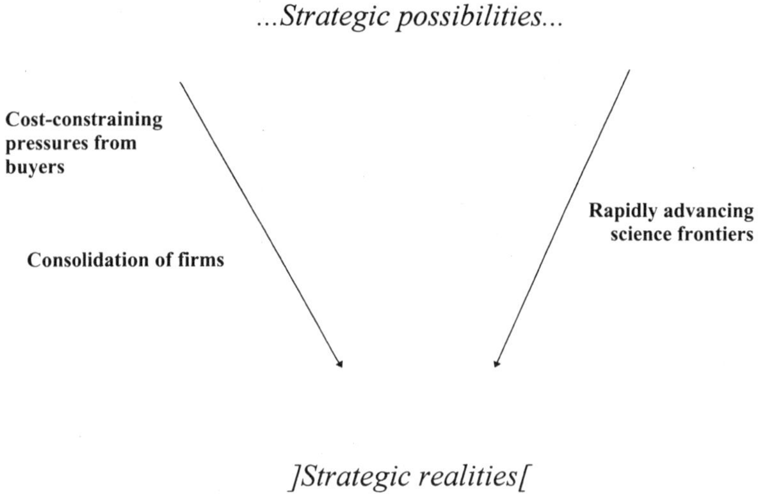

Figure 2. Reduced degrees of freedom.

OBJECTIVE AND ORGANIZING MODEL

The actual or potential gap between scientist and nonscientist with regard to strategy informs both my objective and my choice of model for achieving that objective.

First, what is a model? A model is no more, but no less, than a conceptual tool guiding thought and action. Because it guides thought and action, it is important to match the model (the tool) both to the objectives and to the conditions under which it will be used.

What is my objective? I want to help bridge the gap *by changing the way scientist and nonscientist think about strategy, environment, and technology.* To do so, I have chosen as my tool the organizing model of trajectories.[2] "Trajectory" conjures up images of endpoint or target, energy source, speed, coordinates, space—images quite different from what might be conjured up by the model (tool) of "strategic plan." (An organizing model also provides a common language by which scientist and nonscientist can communicate about strategy.)

Such an organizing model suits my objective—to change the way people think—and suits the conditions under which it will be used. For what is the environment of the biomedical industry if not turbulent? Is not the future likely to contain surprises, perhaps hostile forces impinging on the organization, and myriad possibilities? What better model, then, for the desired future development of the firm than a *trajectory through turbulent space–time*?

Finally, this organizing model emphasizes that many of the capabilities required for the biomedical firm to survive and succeed in a turbulent future are analogous to "rocket science." Deciding on a strategic target perhaps 20 years away, determining and setting coordinates, and ensuring sufficient, effective energy sources to reach the target are not trivial competencies. But, they are exactly the competencies a turbulent future demands.

Technology Strategy

Because my objective is to provoke a novel way of thinking about strategy from the perspective of biomedical science and technology, this book does not focus on strategy per se but on *technology strategy.* My interest is principally in the trajectory defined by overall R&D as well as program- and project-specific decisions. In this organizing model, every discovery effort, every program, every project can be depicted by a vector. The sum of these individual vectors is the R&D vector. Over time, the R&D vectors describe a series of points that (when joined) defines the technology trajectory (Figure 3). Technology strategy is the blueprint for the technology trajectory, just as corporate strategy is the

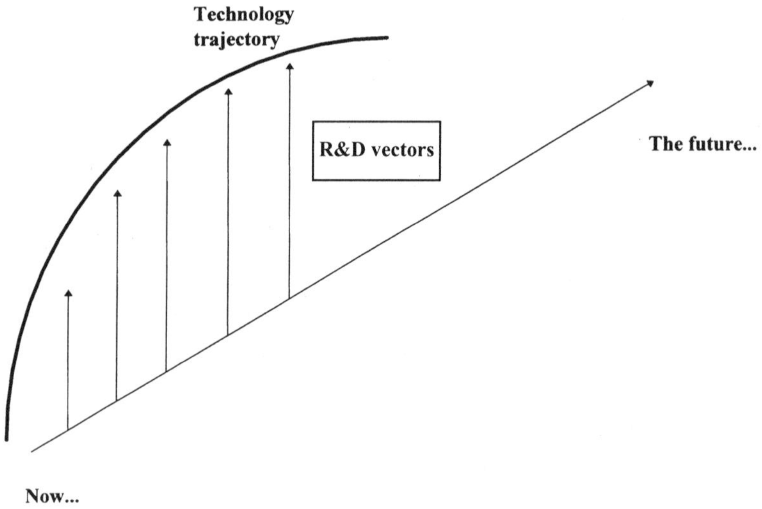

Figure 3. Technology trajectory.

blueprint for the business trajectory. (This is addressed again in the next chapter.)

In effect, the model of trajectories turns the biomedical firm inside-out. The most uncertain work of the organization, research and development, is made paramount. I am not thereby dismissing other functions but rather considering them as infrastructure. This is as it should be. *The most attention must be paid to that which is most uncertain,* both external to the firm and within it.

Turning the firm inside-out ensures that strategy is considered from the perspective of biomedical science and technology. Deciding upon a strategic vision or target requires, in this model, that the leadership also select a technology identity. As defined in Chapter Two and discussed throughout the book, this is my shorthand term for an explicit decision about the general orientation of the firm towards science and technology (such as either defining or remaining on the leading edge of the relevant frontiers). When resource allocation within R&D and all infrastructure functions is consistent with strategic vision and technology identity, the firm's business and technology trajectories are aligned. The gap has been bridged.

Recurrent Themes

I have deliberately chosen a specific objective (to change the way people think about strategy, environment, and R&D) and a narrow focus—the perspective of biomedical science and technology. Objective and focus determine emphasis. The reader will discover that I am more concerned with providing new metaphors, new language, for scientist and nonscientist to use in discussing strategy than with the details of the discussion. I do not, for example, prescribe certain types of portfolio analyses or strategic intelligence activities, but I assume they occur.

The reader will discover that I am more concerned with encouraging an intense feedback loop of communication among corporate, R&D, program, and project managers than with the actual structure of the firm. I contend that, if the interpersonal relationships and communication are effective, structure is secondary.

The reader will also discover that I am more concerned with best principles than with best practices. I expect everyone in the firm to know what relevant best practices are, because they define the competitive floor in each sector of the industry. But, I have based this book on what I believe (from study as well as from experience) are the best principles that should inform strategic decisions.

Finally, given my objective—to help bridge the gap between corporate and technology strategies—the reader will discover that I am more concerned with internal consistency between firm-specific strategic vision and R&D resource allocation than with industry trends. I assume that strategic intelligence activities within the firm ensure that salient industry issues are always considered by the strategic leadership. But, I cannot overemphasize the importance of consistency among strategic vision, technology identity, and scientific and technologic capabilities. I discuss merger, acquisition, and strategic alliances only in the context of ensuring such consistency.

A BRIEF GUIDE TO THE CONTENTS

The preceding discussion of thinking, metaphors, principles, consistency, and so on should not mislead the reader that this is a theoretical as opposed to practical guide to strategy. There is much in the way of "how to" (and "how not to"), with pragmatic illustrations from the industry. There is also conceptual material, especially on technology. The contents, in fact, have been determined by what I have observed from my consulting are some of the major, complex problems facing decision makers.

The book is effectively divided into two parts. Chapters One through Four constitute the first, preparatory portion. Chapter Two provides more information on the organizing model of strategic trajectories that underlies the entire book. I also discuss strategic leadership, what I believe is the appropriate level of strategic sophistication, and vision and values.

Chapter Three presents my methodology for analyzing the external environment of the biomedical organization in accord with the metaphor of turbulent space. Ways to assess levels of turbulence, in order to focus people's attention, and to make use of weak signals are described. Also, the expected impact and unpredictability of crucial events for the biomedical firm are used to define a matrix of generic technology strategies.

Chapter Four concludes the preparatory section of the book and initiates the pragmatic second part: "how to" go about determining, setting, and adjusting vector coordinates by means of R&D resource allocation decisions. Chapter Four also addresses the first complex and difficult problem that inspired this book: how to map science and technology investment to the technology life cycle. The chapter describes three properties associated with technology. With a better understanding of these properties, decision makers can avoid the traps of overinvesting in mature technologies and underinvesting in emerging ones.

Chapters Five through Nine stand on the foundation of the prior chapters. Based on the organizing model of trajectories, Chapter Five addresses a second problem: how to codify the competitive position of R&D. The chapter presents my framework and methodology for evaluating the competitive position of the firm, from the perspective of biomedical science and technology.

Chapter Six takes the framework and methodology a step further and addresses a third problem: how to gauge the internal consistency of science and technology resource allocation. Although it is important to assess the competitive position of R&D, it is even more crucial to assess the consistency among R&D investments, the strategic vision, and the technology identity. As I said earlier, it is only when resource allocation within R&D and all infrastructure functions is consistent with strategic vision and technology identity that the firm's business and technology trajectories are aligned.

Chapter Seven takes as given that the leadership are (now) capable of achieving internal consistency, as above, and addresses a fourth problem: how to determine the consequences of changing competitive position. A practical guideline and illustrations are described, assuming that the change is necessary for the biomedical firm's survival.

Chapter Eight addresses a fifth problem: how to assess the fit of external knowledge sources with the firm's vision and technology identity. Because of rapidly advancing science frontiers in the biomedical industry, no firm can remain proficient in core technologies, over the long

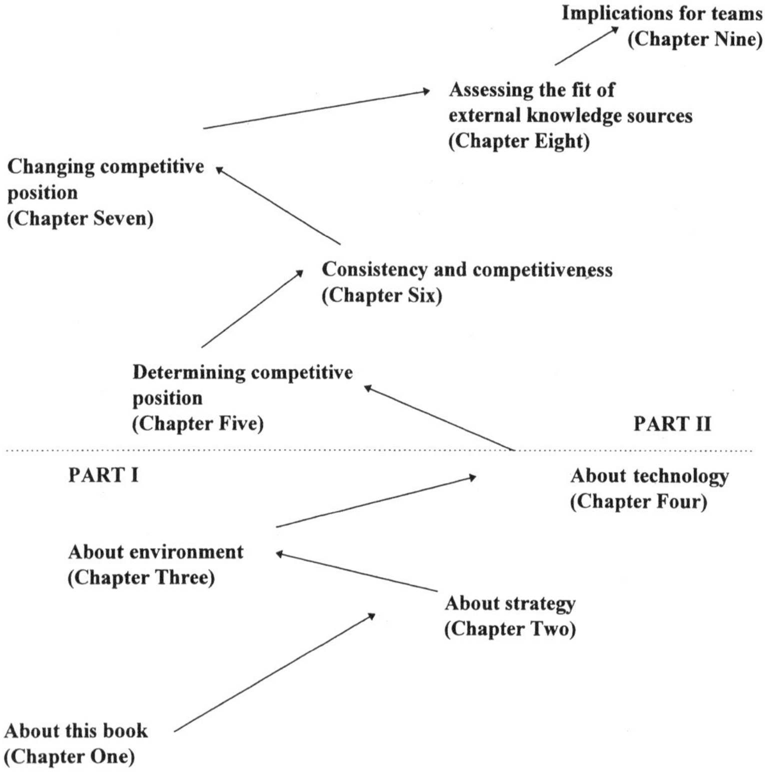

Figure 4. A guide to the contents.

term, by itself. The chapter provides criteria to examine this fit (making use also of mergers, acquisitions, and strategic alliances), so that the strategic leadership can make competitively advantageous science and technology investments.

An entire book could be written on the topic of the final chapter—implications of competitive technology strategy for R&D teams. Very briefly, this chapter addresses a sixth problem: implementation. In Chapter Nine, I try to make clear that the process of determining, setting, and adjusting the coordinates of the R&D vectors becomes actionable at program and project levels. *R&D teams implement competitive technology strategy.*

Figure 4 summarizes the contents of the book.

NOTES

1. The need to link corporate strategy with R&D was described by P. A. Roussel, K. N. Saad, and T. J. Erikson in *Third Generation R&D: Managing the Link to Corporate Strategy,* Harvard Business School Press, Boston, MA, 1991. Other pertinent authors who focus on the gap between corporate and technology strategies are listed below:

   R. Coombs and A. Richards, "Technologies, Products and Firms' Strategies, Part 1—A Framework for Analysis," *Technology Analysis and Strategic Management, 3,* 1991.

   A. C. Perrino and J. W. Tipping, "Global Management of Technology," *Technology Analysis and Strategic Management, 3,* 1991.

   R. Coombs and A. Richards, "Strategic Control of Technology in Diversified Companies With Decentralized R&D," *Technology Analysis and Strategic Management, 5,* 1993.

2. The term "trajectory" is not unique to this book. As catalogued by Paul Adler ("Technology Strategy: A Guide to the Literatures," unpublished report, Department of Engineering and Engineering Management, Stanford University, 1988), a number of scholars have described "development paths" as "trajectories." Some of the earliest examples include:

R. R. Nelson and S. G. Winter, "In Search of a Useful Theory of Innovations," *Research Policy,* January 1977.

D. Sahal, *Patterns of Technological Innovation,* Addison-Wesley, Reading, MA, 1981.

G. Dosi, "Technological Paradigms and Technological Trajectories," *Research Policy, 11,* 1982.

Adler stated that "business strategy has its parallel in a technology strategy, which identifies a vector characterizing the projected development path of the organization's technology efforts and capabilities" (p. 9).

CHAPTER TWO

ABOUT STRATEGY

The purpose of this chapter is to describe in more detail the model of, and approach to, strategy. Using the organizing model of trajectories, the biomedical organization is turned "inside out." Research and development are treated as functions central to the strategic process and R&D programs and projects as pivotal to strategic success. In addition to discussing the model, this chapter discusses two key concepts—strategic trajectories and alignment of trajectories—reviews some characteristics of strategic leadership and levels of strategic sophistication, and discusses strategic vision and organizational values implied by that vision.

THE ORGANIZING MODEL

In a simple, stable, certain environment with (almost) perfect information, the model of *strategic plan* is appropriate. The word "plan" comes from the Latin for "level ground."

When people in the organization can with some confidence "see ahead," a plan is a satisfactory conceptual tool. But when the future will bring unknowns, unexpected events, and discontinuities (i.e., uncertainty and imperfect information), a more complicated tool is needed, in line with the principle of requisite variety.[1]

As I said in Chapter One, this book is not about strategic planning, although the words "strategy" and "plan" appear throughout. Rather, it is about the more complicated but more appropriate to the times activities of *determining, setting,* and *continuously adjusting* the coordinates of the R&D vector. Over time, these vectors define the technology trajectory.

A trajectory is the path described by an object moving through space and time because, and only because, energy is supplied. This path can be deliberately and rigorously calculated, but the trajectory is by no means fixed. Those who send objects into space expect all manner of forces to disturb the trajectories. They must determine, set, and then *continuously adjust* coordinates, in response to feedback on the object's deviation from the path. They may also have to recalculate a trajectory if the deviation is large, albeit keeping the object still aimed at the chosen destination.

According to this book's organizing model, the biomedical organization is the object, the turbulent health care environment now and into the twenty-first century is the relevant space and time, and energy is supplied by research and development. The term strategic trajectories refers to the metaphoric paths whose blueprints are (1) corporate strategy (which guides the allocation of business resources and essentially defines the organization's *business trajectory*), and (2) technology strategy (which guides the allocation of science and technology resources and essentially defines the organization's *technology trajectory*).

Moreover:

- The most uncertain work of the biomedical organization—research and development—is central to the strategic process.
- Programs and projects are the booster rockets that provide energy and keep the business and technology trajectories aligned.
- Success of the biomedical firm in the twenty-first century depends upon the leadership's (1) ability to create an instrumental technology strategy providing superior competitive advantage, (2) sophisticated understanding of technology and the strategic significance of programs and projects, and (3) skill in enhancing technology strategy continually, by astute allocation of science and technology resources.

ALIGNMENT OF STRATEGIC TRAJECTORIES

Although the focus of this book is the technology trajectory, my objective is to help bridge the gap between corporate and technology strategies. Thus, it is important to delineate the relationship between them. "Bridging the gap" is another way of stating that "business and technology trajectories are aligned."

The strategic vision is the target or endpoint of corporate and technology strategies. Coordinates of the business vector are defined by corporate strategy, which provides the framework for decisions about (among others): business objectives, the business(es) in which the biomedical firm will compete, the general types of products that will be offered, the markets in which they will be sold, and so on.

These decisions guide the allocation of business resources. As I will discuss in Chapter Four, the latter decisions also specify the firm's *domain*.

In deciding upon a strategic vision, corporate and R&D managers must also select a *technology identity*. Discussed again in Chapter Six, this is my shorthand term for the values about science and technology held by people within the organization. Based on these values, technology identity reveals itself as the desire to *define* the leading edge of relevant sciences and technologies, or to *remain on* that leading edge, or to *pursue* it. One such general orientation always characterizes a firm. My position throughout this book is that the choice must be made explicit.

Given strategic vision and technology identity, coordinates of the R&D vector are defined by technology strategy, which provides the framework for decisions about (among others): science and technology objectives, core technologies in which the firm will be proficient, general technologies that will support R&D, and so on. These decisions guide the allocation of science and technology resources (people, time, facilities, equipment, money) to discovery programs and development projects. In other words, technology strategy is the context for decisions about the types of programs and projects to be funded, and how; their priority ranking within the portfolio; and the criteria to be used in making tradeoffs.

At the same time, events occurring at the program and project levels, such as scientific breakthroughs or dead ends, or new information on competitors' activities, must inform decisions to increase, decrease, or shift personnel and funding levels. It is in this way that booster rockets are deployed or changed, to adjust alignment of business and technology trajectories and/or to enhance technology strategy. Alignment of strategic trajectories, and enhancement of technology strategy, demands an intense feedback loop of two-way communication among all spheres of management: corporate, research, development, program, and project. Strategy is, literally, "embedded in projects"; thus, corporate, R&D, program, and project decisions are inextricably linked.[2]

Of course, science and technology resources are not just used in R&D. The biomedical firm's scientific and technologic capacity must continually be developed. As the frontiers of science and technology advance, the intellectual capability of R&D must advance as well, by incorporating new technologies, by training and skills improvement of current staff, and by recruiting and hiring new staff. Competitive technology strategy demands another feedback loop between science and technology *use* and science and technology *advancement*.

Figure 5 summarizes some of the more important decisions guided by a competitive technology strategy. Figure 6 illustrates the feedback loops of communication required to keep the business and technology trajectories aligned.

Example: Lucida NeuroSciences

Lucida NeuroSciences, a hypothetical midsize pharmaceuticals company, provides a practical illustration of the above issues. (The company is also discussed in Chapters Five, Six, and Seven.)

Because Lucida is too small to compete across a broad range of therapeutic areas, R&D leadership propose, and corporate management agree, on a niche strategy with an emphasis on central nervous system therapies. (Two of their scientists had discovered a novel mechanism to reverse neurodegeneration *in vitro* and in appropriate animal models.) The leadership's vision for Lucida is to be always among the top three firms by market share in the niches in

Spheres of Management Action	Spheres of Technology Action	
	Use	**Advance**
Corporate	With R&D, defines unique technology identity; sets R&D budget	With R&D, changes identity and budget as appropriate (where? when? how? based on?)
Research	Selects discovery programs, research projects; allocates resources; assesses new information and data; communicates information	Changes priorities and allocation (where? when? how? based on?)
Development	Selects projects; allocates resources; assesses new information and data; communicates information	Changes priorites and allocation (where? when? how? based on?)
Program and Project	Determines concept viability; carries out efforts; seeks and interprets new data; communicates information	Changes decision criteria (where? how? when? based on?)

Figure 5. Competitive technology strategy.

which they choose to compete. The firm's technology identity is to *define the leading edge* of relevant sciences and technologies, and a major business objective is to introduce a breakthrough therapy within eight years.

In this context, the following types of decisions must be made:

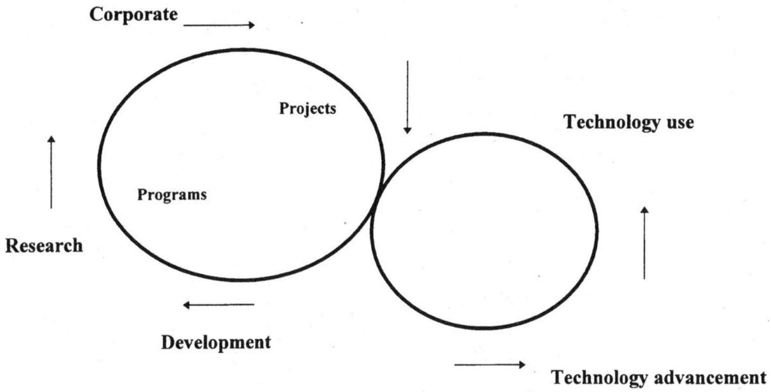

Figure 6. Keeping trajectories aligned.

- In the near term, how many people and how much money should be allocated to discovery and in what areas, including target identification; of this amount, how much should be allocated to in-house versus outside partners, and on what basis?
- Given current and expected competition and consolidation within the industry (see Chapter Three), what criteria should determine which potential candidates from discovery are put forward into preclinical research and then clinical development, and should those functions be accomplished solely in-house, with a partner, or by means of contract organizations (and on what basis)?
- In the medium and long term, how many people and how much money should be allocated to efforts devoted to *breakthrough* innovations versus *incremental* advances?
- Over the long term, what science and technology resources (people, money, facilities, time, equipment) will be needed to ensure that R&D continues to define

the leading edge, consistent with their technology identity? How many and what types of people should be recruited, and what types of training should be provided to current staff? What partnerships should be considered (academic or industrial laboratories), etc.?

On the basis of these and related science and technology decisions, marketing managers must decide what resources will be needed to market the products to large managed care organizations when the products are approved; and should marketing be undertaken by in-house personnel alone or with partners. They must also determine what marketing capability will be required to support products over their expected life. At the same time, manufacturing and engineering managers must decide what equipment and capacity will be required and when and where (in-house or contracted out) the investments should be made, so that plants are available in advance of the competition. Managers in all infrastructure departments—quality, regulatory affairs, finance, etc.—must make similar decisions.

The above process represents the ideal. Because of the gap in understanding between scientist and nonscientist, a more likely situation is one in which the leadership, responding to price and utilization controls on the biomedical industry (Chapter Three), revise corporate strategy. They presume that current activities in R&D, such as amount and timing of resource allocation and project decision criteria, are reasonably consistent with the new business objectives. Program and project managers are aware of changes in corporate strategy but not sure of the implications for the portfolio; nor are they sure of the impact of their own decisions on the attainment (or not) of business objectives. In any case, communication within the organization is primarily top-down, challenging of assumptions is discouraged, and information systems appear to be designed to provide only good news. The case of Solar Imaging will illustrate. (This company is also discussed in Chapters Five and Six.)

Example: Solar Imaging

A few years earlier, corporate leaders of this hypothetical midsize medical device firm had announced aggressive growth, profitability, and market share goals. In accord with new business objectives, they shifted overall resources. Their plan was that, in 3 years, one division (Division X)—which had a large market share but whose products were in danger of obsolescence—would receive about 30% of total R&D, down from a peak of nearly 80%. A second division (Division Y), which accounted for only 2% of current company sales but whose pipeline included several radical innovations, would receive about 70% of total R&D.

R&D managers in Division X, however, overinvest in their mature technology, using external alliances to make up for the decreasing corporate budget. Project managers are overwhelmed by too many apparently equivalent product prototypes (one consequence of a mature technology.) They have no context in which to decide on priorities or to make tradeoffs. Discouraged by this state of affairs, a number of key scientists leave.

In Division Y, R&D managers enthusiastically support the changed budget. However, they underinvest (in comparison with rival firms) in capabilities to support their radical innovations and realize too late that a rival is likely to beat them to the market by 12 to 18 months. If communication within Solar Imaging were two-way, R&D directors would have learned from their project managers that the portfolio was too broad and that top-priority products were underfunded.

At the present time, business objectives are far from being achieved. Because of its cash position (essentially the result of underfunding R&D), Solar Imaging is considered an attractive acquisition candidate.

STRATEGIC LEADERSHIP

The organizing model of trajectories requires a new model of leading as well. *Strategic leadership* is neither a corollary of title and position nor confined to people at the "top" of R&D. That model of leading is consistent with the model of *strategic planning*. In the new model, strategic leadership requires (in addition to appropriate scientific and technical competency) the ability to envision, inspire, focus, challenge, and communicate.[3] Moreover, this ability must be widely dispersed throughout R&D, or the biomedical firm will not remain competitive over the long term.

The "discipline" underlying technology strategy is "rocket science." That is why strategic leadership—determining, setting, and adjusting the coordinates of the R&D vectors—is a most important, terrifying, difficult, and wonderful task.

It is most important and most terrifying for the same reason: The fate of the organization is in the hands of those making science and technology resource allocation decisions. Determining, setting, and adjusting coordinates puts the biomedical firm on a course of success or failure. In a turbulent and hostile environment, there is little forgiveness for being "almost right."

It is a most difficult task because of the complexity and

volatility of the health care environment. Every industry faces the latter characteristics to some degree, but the biomedical industry also faces:

- heavy pressure for economic success, because of the industry's geopolitical significance
- rapidly advancing science frontiers, with technology, process, and product life cycles that are being compressed in large measure because of these rapid advances
- increasing constraints on product revenue
- industry consolidation, resulting in even more intense competition.

In short, the biomedical industry confronts a greater degree of complexity and volatility than most industries. As I will explain in Chapter Three, the environment of the biomedical industry can accurately be defined as *turbulent*.

Finally, strategic leadership is a most wonderful task for the same reasons it is important, terrifying, and difficult. It is a genuine challenge, it is intellectually exciting, and it is critical to the viability of the organization. Few jobs can make that claim.

LEVELS OF STRATEGIC SOPHISTICATION

Strategy in general and technology strategy in particular can be approached by one of three levels of sophistication: reacting to, positioning within, or striving to shape the environment.[4] I am convinced that only the last can produce a viable and successful biomedical firm.

Reacting

The lowest level of strategic sophistication consists of *reacting* to the apparently random events, trends, issues in the environment (Figure 7). Without explicit decisions, intense efforts, and commitment to move beyond this level of sophistication, the leadership will only achieve what has been described as minimalist strategy and performance. A reactive or minimalist technology strategy is no more than a series of short-term decisions to unplanned R&D vectors, with the hope of achieving coherence of business and technology trajectories. Given the turbulence of the health care environment, that hope is very slim.

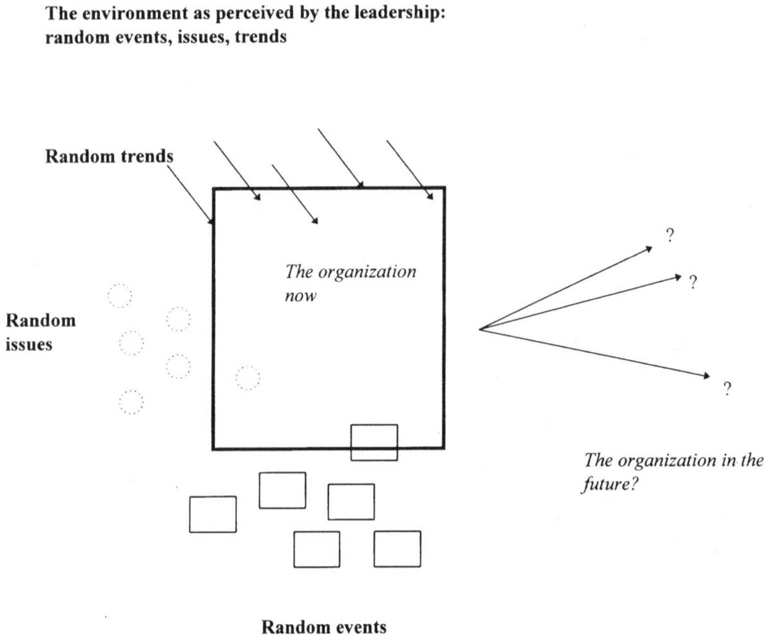

Figure 7. Reactive strategy.

Positioning

A higher level of sophistication consists of *positioning* the biomedical firm within the environment in terms of the familiar SWOT paradigm (**s**trengths, **w**eaknesses, **o**pportunities, **t**hreats) (Figure 8). Positioning can result in satisfactory although not outstanding performance, because the environment is taken as given and coordinates of the R&D vector are set within those constraints. Accordingly, the technology trajectory will be limited by what people perceive as the "box" circumscribing the organization.

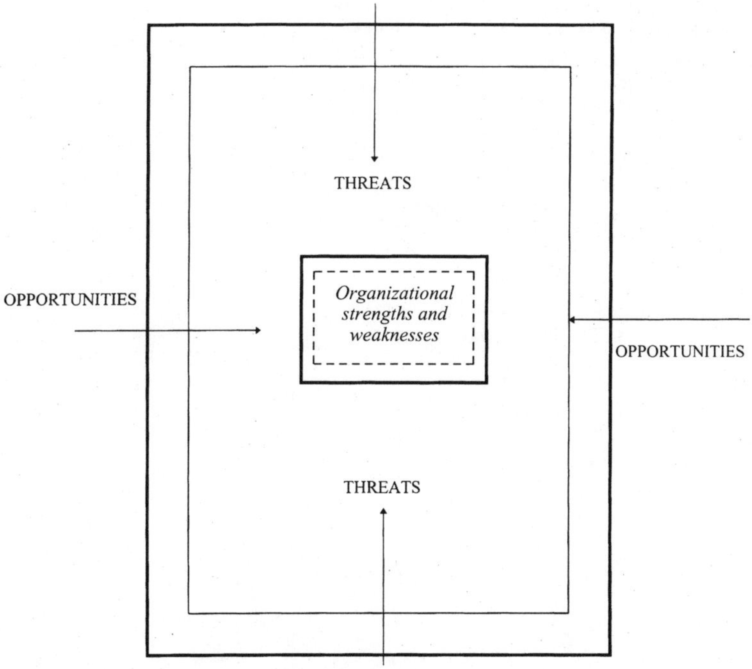

Figure 8. Positioning strategy.

Shaping

For the organization to be truly successful, I believe the strategic leadership must strive for the highest level of sophistication, which is to be *instrumental* in shaping the environment and the future. Being instrumental means not taking the environment as given, as the source of constraints, but as the source of raw material to be shaped by creative science and technology decisions. In other words, the leadership must envision what might be and then actively work to achieve it (Figure 9).

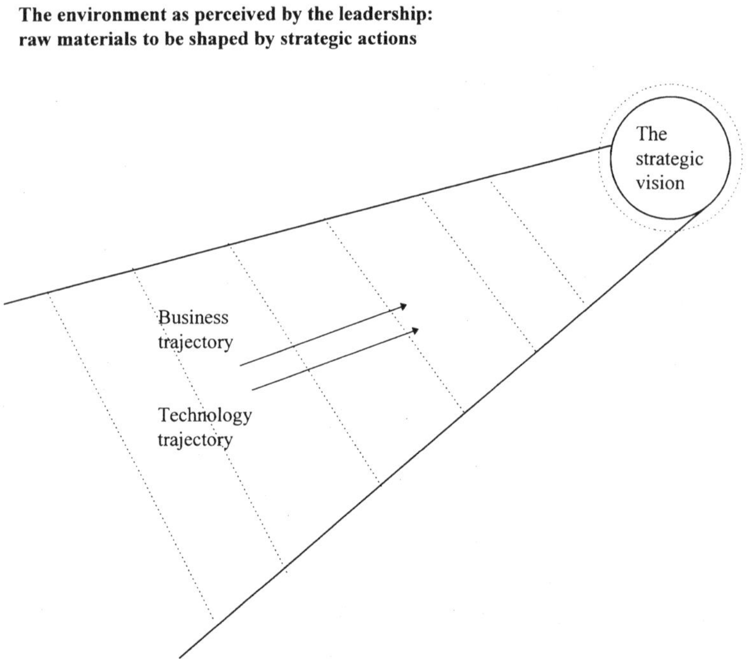

Figure 9. Instrumental strategy.

Much has been made of fitting the organization to the external environment, and the environment is certainly influential (Chapter Three is devoted to it, in fact). But, being instrumental requires that people are primarily concerned with what the organization will be; in other words, with its unique character and identity.[5] What the organization will do must follow from the determined identity (the "being"), rather than vice versa (Figure 10).

Because organizational identity is not achieved quickly,

BEING

Human character	Treatment of employees, customers, other constituencies
Scientific and technologic competence (technology identity)	State of the art, traditional, posthumous
Geographic scope	Global, multinational, national
Competitive position	Leader, middle of the pack, follower
Contribution to communities	Model organization, good citizen, follow letter of law

DOING

- What businesses?
- What products?
- What programs?
- What projects?
- What resources?
- What collaborations?
- What alliances?

Figure 10. Strategy: being and doing.

the period of time with which instrumental strategy must be concerned is long—one or two decades, although tactics (the "doing") will have to be adjusted continually. Being instrumental carries the highest risks, of course, but it is the only approach that can result in outstanding performance.

From the perspective of biomedical science and technology, at the instrumental level of sophistication the strategic leadership of R&D will:

- take as their endpoint the strategic vision and technology identity
- continually conduct analyses of the turbulent space through which the technology trajectory will pass over time, considering what can be done to shape the raw materials (i.e., shape the future)
- decide on tactics, given the results of these analyses
- ensure that trial actions are taken, feedback is provided, performance is evaluated, and trial actions are taken, etc.

As implied by the above, instrumental technology strategy comprises a meta-loop of activities that continue as long as the biomedical firm endures. They are the activities that set and reset the coordinates of the R&D vectors and ensure that the firm stays faithful to the ideals described by strategic vision and technology identity.

VISION AND VALUES

Although they may appear to be anomalous topics in the new model of strategy, in fact both *vision* and *values* are of paramount importance (see also Chapter Six).

The Strategic Vision

Deciding upon the endpoint of the business and technology trajectories, the crucial first step of being instrumental, is too important to be trivialized, to be assigned to staff, or to be left to consultants. No matter how excellent the staff or the consultants, only the strategic leadership can and should finally articulate the vision. Consultants can conduct analyses and help facilitate the process; staff must contribute; but the leadership must accept the final responsibility for what the vision describes.

There are a number of excellent books and articles about genuine "strategic visioning," that converge on the following:

> The strategic vision consists of shared agreement on what the organization could be. It presents a "clear image of what the organization can and should become if it is to realize its full potential."[6]
>
> The strategic vision is *not* a description of what the organization will or should do in response to the environment, nor is it an ambiguous statement that can be applied to practically any organization, nor is it just a "statement of some sort printed on an elegantly framed plaque hanging in the lobby."[7]

The right vision:

- *Serves as an energy source for all stakeholders.* When people reflect on the vision, they should want to work for its achievement. The vision should be truly "motivating."
- *Provides the context, touchstone, and framework within which to analyze the environment.* Every biomedical organization faces the same environment. In the suc-

cessful firm, people perceive that environment in light of their shared vision and technology identity. They see opportunities that others miss, experiment when others maintain course, and persist and succeed in what others have given up as impossibilities.

- *Helps people "organize and focus internal resources" on that which is most important to the organization.* There are many options involved in allocating science and technology resources. The strategic vision helps to make the appropriate choice plainer to everyone.
- *Serves as the "basis for every important decision."* Given a shared vision, people can examine whether the decision they are about to make moves the biomedical firm closer to, or further from, the desired end. Without a vision, decision making sends the organization on a journey without a destination.[8]

The right strategic vision must be ambitious; it must stretch people in the organization and lift them out of the usual and customary; it must be inspired and inspiring (Figure 11). Some excerpts from actual biomedical organizations will illustrate:

> We will always be in the top five global . . . companies. . . . No matter what function is examined, our solutions will meet genuine needs in an innovative way. Our competition will wonder how we manage to be so successful. Our constituencies will respect us without reserve; our employees will be the best in the business and will love to work here.
>
> Our firm will be first in every niche in which we compete. Our processes will set the standards for the competition, and our name will be synonymous with quality. . . . All of us will stand accountable for our

- Consists of shared agreement on what the biomedical firm could be
- Is a source of energy for all stakeholders
- Serves as the context, touchstone, and framework for analyzing environment
- Provides a focus and organizer of action
- Is the basis for every important decision
- Must be ambitious
- Must be inspired
- Must be inspiring

Based on T. Nolan, L. Goodstein, and J. W. Pfeiffer, *Plan or Die*, Pfeiffer and Co., San Diego, CA (1993).

Figure 11. The right stategic vision.

> work and determined to improve people's health in meaningful ways.
>
> We will be a world-renowned organization providing expert consultative services and operational support in the area of health care research. We will take a leadership role in applying and developing state-of-the-art technologies and methods, to ensure that our studies meet the highest standards of investigative, therapeutic, and regulatory practices. Scholars of the highest caliber will want to work with us. . . .

The strategic vision also serves as the starting point for a discussion of organization values.

Organizational Values

It is impossible to articulate a truly inspired strategic vision without confronting the notion of what *values* this vision implies and requires. In fact, if the process of defining the vision is effective, the subject of values must emerge, although they may be termed behaviors, norms, culture, etc.

The following quotation on the topic of values should be helpful:[9]

> Developing an explicit statement of values often proves to be a useful activity for a management team. It forces the team members to confront their own personal values and to determine the degree to which they are shared by others. . . .
>
> Values have always implicitly driven the decisions of the organization. Making them explicit provides for a more conscious choice and allows managers to be directly guided by these values as well as to question them.

For any biomedical firm, an ambitious strategic vision requires, at a minimum, the values of risk willingness, flexibility, speed, innovation, candor, and collaboration within R&D. If these are not currently the values, then the new model of strategy requires that this issue be addressed first.[10] There is no point in attempting to determine, set, or adjust coordinates of the R&D vectors unless vision and values are aligned. I take this up again in Chapters Six and Seven.

BEST PRINCIPLES

As I said in Chapter One, this book is not informed solely by the *best practices* in industry but, more importantly, by what I believe are the *best principles* that should underlie all efforts in the realm of strategy. There is nothing wrong with best practices. It is important to know what they are, because they define the biomedical firm's competitive floor.

But I contend that people should not expend a lot of energy benchmarking against best practices. Consider this: "Best practices" is a retrospective judgment. The designation is based on one or more organizations' successes, which stem from what people in the organization did in the past. Best principles, on the other hand, are prospective in nature. They are not based on what was, or what is, but on what should be (and they emerge from thoughtful reflection on relevant concepts or theories plus experience).

When an organization has caught up with best practices, it has only caught up with the noninnovating firms. Better to benchmark progress against the best principles I propose in the following chapters; this should enable the firm to leapfrog even companies known (in the past) for innovation.

NOTES

1. The principle of requisite variety was described by W. R. Ashby in *Introduction to Cybernetics,* London, Chapman and Hall, 1956. Ashby called it the law of requisite variety, and derived it from a type of game theory involving two players, "R" and "D."

   > Let V_D be the variety of D, V_R that of R, and V_O that of the outcome (all measured logarithmically). Then . . . V_O's minimum is $V_D - V_R$. If V_D is given and fixed, $V_D - V_R$ can be lessened only by a corresponding increase in V_R. Thus, the variety in the outcomes, if minimal, can be decreased further only by a corresponding increase in that of R (p. 207).

   See also S. Beer, *Brain of the Firm,* Wiley, New York, 1981.
2. Strategy as embedded in projects is discussed again in Chapter Nine. The quotation is from Paul Adler, "Technology Strategy: A Guide to the Literatures," unpublished report, Department of Engi-

neering and Engineering Management, Stanford University, 1988, p. 36.

3. The reader will note that these characteristics of a leader were described by Peter Senge in (for example) "The Leader's New Work: Building Learning Organizations," *Sloan Management Review, 13,* 1990.
4. Paul Adler, as above, is the inspiration for the three terms describing levels of strategic sophistication. He described minimalist strategy as the "simple recognition of necessity"; a higher level of "strategic development would consist of a framework for deciding in which subregions of these environments the firm might optimally position itself." The highest level "would acknowledge that the firm may have the possibility of shaping the environment and creating new opportunities; advancing the technology, influencing values and government policies, changing market demand, and altering the terms of competitive rivalry" (all cited on p. 10).
5. The notion of "identity" as an integral part of strategy was described by Kenneth Andrews in *The Concept of Corporate Strategy,* Dow-Jones-Irwin, Homewood, IL, 1980.
6. This and other quotations on vision (cited below) are taken from p. 31 of a very readable paperback: *Plan or Die,* by Timothy Nolan, Leonard Goodstein, and J. William Pfeiffer, Pfeiffer and Co., San Diego, CA, 1993.
7. A similar perspective on vision and values is discussed in "The Vision Thing," by Chris Lee, *Training* (February 1993), pp. 25–32. The quotation is from p. 26.
8. These points come from *Plan or Die,* pp. 31, 32, and 32, respectively.
9. Ibid., p. 41.
10. Alice M. Sapienza, *Managing Scientists,* Wiley, New York, 1995.

CHAPTER THREE

ABOUT ENVIRONMENT

The purpose of this chapter is to review the necessary and important mechanics of analyzing the biomedical industry environment. A unique three-part methodology is proposed for examining the environment, or "space," in the new model of strategy. First, key components of each dimension constituting this space are identified. Second, the level of turbulence of each dimension is assessed and weak signals of change are defined. Third, the interdependence among the dimensions is determined and the impact and unpredictability of events surfaced by the prior analyses are noted in a four-cell matrix. From this unpredictability-impact matrix, generic technology strategies appropriate to the environment of the biomedical firm can be derived.

ENVIRONMENT DEFINED

In the new model of strategy, environment is the space through which the strategic trajectories will pass. It consists of both forces with which the biomedical firm must contend (in the manner of meteor showers and other hazards) and raw materials to be shaped in the achievement of the strategic vision.

As I stated in the prior chapter, outstanding organizational performance depends, among other things, on peoples' *not* being limited in their thinking by models or best practices in which environment is treated as a set of constraints (the box in Figure 8). By approaching environment as space–time, as forces acting on the strategic trajectories and as raw materials to be shaped, the analytic focus is changed *from* boundaries and limits *to* possibilities. When the biomedical firm is not perceived as passive in the face of givens but as active in the realm of potentials, the outcome is greater strategic creativity, even when the degrees of strategic freedom are reduced (I address this in more detail in Chapters Five and Six).

In this chapter, I propose a unique three-part methodology for analyzing environment as space–time. The first part involves disaggregating this space into seven dimensions, to identify key components for the biomedical firm over a 10- to 20-year period.[1] The objective of this part of the methodology is to allow people to analyze what is otherwise a chaotic surround in a systematic and rigorous manner.

The dimensions are not equally important at any given time. Because of human cognitive limits and imperfect information, the strategic leadership must focus people's attention on those dimensions that will be, in their informed judgment, *most turbulent.* My reasoning is this: There are two possible consequences of turbulence. First, the biomedical firm is more likely to be knocked off course under conditions of turbulence. Second, turbulent dimensions also provide the stuff, the raw materials, for shaping the environment. Focusing on the most turbulent dimensions prepares people in the organization to prevent the first consequence and to take competitive advantage of the second.

Thus, the second part of the methodology consists of assessing levels of turbulence and identifying appropriate weak signals to be monitored, dimension by dimension. "Turbulence" is defined as a condition in which the rate of change is fast, the magnitude of change is large, and the unpredictability of change is high.[2] The objective of this part of the methodology is provide a framework for people in the firm to come to shared agreement on the levels of these parameters (rate, magnitude, and unpredictability). Based on their agreement, the dimensions can be ranked from most to least turbulent, thus sharpening strategic focus.

The concluding part of the methodology has two steps. The first involves determining the *degree of interdependence* among the dimensions. Up to now, each dimension has been considered as if it were a discrete force. When this metaphoric space is reexamined as a whole, when the interdependence among dimensions is considered, the priorities about where to focus are likely to shift. People may find that a dimension first judged to be stable is, when seen in context, actually turbulent because of its interdependence with other more turbulent dimensions. The objective of this step is to help the leadership decide if a reranking of dimensions is warranted—again, to focus strategic attention.

The second step involves selecting the most important issues and events surfaced by all prior analyses and coming to some judgment as to the impact of these issues and events on the biomedical firm (low or high) and their perceived unpredictability (low or high). I have delineated four generic technology strategies, based on the four-cell matrix derived from impact and unpredictability judgments. The objective of this step is to ready the biomedical firm for action at a time when the cost of responding is low and the probability of preempting rivals is still high.

Before continuing, I want to emphasize that environmental scanning is everyone's responsibility. Unless each person contributes to strategic intelligence, science and technology decisions (and, consequently, performance of the biomedical firm) will be less than optimal. The widest scan, after all, requires everyone's participation. Moreover, scanning, intelligence gathering, and analyzing must be ongoing activities supported by two-way communication among corporate, research, development, program, and project management. As I stated in Chapter Two, instrumental technology strategy is a meta-loop of activities consisting of analyses, trial actions, review of feedback, analyses, etc., that continue as long as the biomedical firm endures.

Below, I review the first part of my analytic methodology: identifying key components of the dimensions of space.

PART I: KEY COMPONENTS OF DIMENSIONS

For the purpose of systematic analyses, "space" is disaggregated into seven dimensions, and each is examined as if it were a discrete force acting upon the biomedical firm:

- macroeconomic
- microeconomic
- scientific and technologic
- epidemiologic
- demographic

- political–regulatory
- sociocultural.

Each dimension must be rigorously, exhaustively, and *separately* studied, to identify key components relevant to the biomedical firm and to consider the outcome on the firm of changes in those components over a 10- to 20-year period.[3] I have suggested key components in each dimension only to illustrate the process. These are by no means the only key components, events, and issues to be considered. Also, although my emphasis is on the U.S., in-depth analyses should be completed for every national market in which the biomedical firm does (or expects to do) business.

Macroeconomic Dimension

KEY COMPONENTS: **Investment, Trade, National Budget, Currency, Employment**

The macroeconomic dimension has enormous direct and indirect ("trickle down") consequences for health care in general and the biomedical industry in particular. Some of the key components in the U.S., and their impact on this industry, are briefly discussed below.

Investment, Trade, and National Budget. The Organisation for Economic Cooperation and Development stated that

> . . . the United States is at a serious disadvantage with respect to the rest of the world in terms of its lack of investment and especially saving. . . . The worsening historical trend in Federal indebtedness has not been matched by asset acquisition . . . , [which] does not

augur well for the Federal contribution to future economic growth.[4]

Federal indebtedness has increased in part because of negative balances of trade over the past decade. As a consequence, when more of the budget is spent on the interest resulting from these negative balances, comparatively less is available for, among other things, health care and biomedical research. As less government money is available for health care, price and utilization controls on the health care and biomedical industries (discussed below under "political–regulatory") increase. As less money is available for health-related research, the publicly funded infrastructure of the biomedical industry is jeopardized. This is of special concern in the U.S., which has the largest health R&D effort in the world.

The U.S. spends about 15% of annual GDP on health R&D (which includes biomedical R&D), compared with 3–5% for the research-intensive nations of Germany and Japan.[5] Of the $33 billion spent on health R&D in 1994, the U.S. National Institutes of Health accounted for $10.3 billion, primarily for so-called basic or fundamental research that underlies the biomedical industry.[6] Pharmaceutical and biotechnology companies spent more than that on applied R&D. As many have noted, government, industry, and academia play complementary roles in biomedical R&D; thus, if one agent (e.g., government) were to reduce its investment, all of biomedical R&D could suffer.

Currency. As the value of the U.S. dollar falls (because of mounting budget deficits), the cost of imports increases and so does the trade deficit. The outcome is downward pressure on the amount of government money available for health care and biomedical research.

Employment. As employment shifts to lower-paying jobs because of downsizing (some of which is, in turn, triggered by the weakness of the U.S. dollar), individuals have less money to pay for health insurance or products like drugs. (Other key components would be identified and analyzed as relevant.)

In summary, the macroeconomic dimension presents perhaps the greatest hazard for the biomedical firm's technology trajectory. This dimension contains much that is beyond the strategic leadership's ability to control, even with instrumental sophistication. Determining appropriate weak signals (see below) for this dimension is critically important.

Microeconomic Dimension

KEY COMPONENTS: **Rivals, Suppliers, Buyers, Substitutes, Potential Entrants, Barriers to Entry**

One perspective on the microeconomic dimension (i.e., competition) is provided by the discipline of industrial-organization economics. As described by Michael Porter, the intensity of competition depends on five forces: *rivals*, or organizations providing similar products/services; *suppliers* (of inputs); *buyers* (of outputs); *substitute* products; and *potential entrants* to the market, with the associated barriers to their entry.[7] Although I will use another means of codifying the competitive position of R&D in Chapter Five, this perspective provides important insights into the biomedical firm's competitive context. Each force is reviewed below.

Rivals. Take, for example, the pharmaceuticals sector of the biomedical industry. In this sector, all firms providing medicines for, and conducting R&D in, the same diseases constitute rivals. If there are many, competition will be more intense and more hostile. In this situation, speed-to-market capability, which affects the timing of introduction of new products, is critical to success. Firms that are not among the first two or three to the market may not capture sufficient sales to warrant product introduction. With few rivals, it may be easier to gain market share, assuming the firm is capable of bringing products to market more quickly than the competition.

Suppliers. Suppliers are those organizations providing inputs such as labor, raw materials, scientific expertise, technologies, etc. Strong suppliers (those that must be of concern to the strategic leadership) have enough clout to affect the biomedical firm's profit margins.

For example, if a pharmaceuticals company relied solely on one drug delivery system manufacturer with proprietary technology, it would have little recourse but to pay whatever price was asked. And, every increase in its cost of goods would, all else being equal, decrease the company's profit margins. Other strong suppliers include manufacturers of raw materials that are irreplaceable in the short term. Also, university or government programs in critical disciplines may be strong suppliers (in this case, of expert scientific labor). Maintaining good relationships with these as a source of scientific expertise, and preempting rivals, will add to the cost of doing business in the near term but in the long term will provide competitive advantage. This is discussed again in later chapters.

Buyers. In health care generally, buyers include (1) the patient, who actually consumes the medicines or for whose care the devices or diagnostics are used; (2) the provider (usually the physician), who decides what product or ser-

(usually the physician), who decides what product or service will be consumed; and (3) the insurer, who reimburses the product or service. Strong buyers in the U.S., with the ability to affect profit margins, include managed care organizations (MCOs). Whether biomedical products are used in the hospital or the ambulatory setting, MCOs are demanding more evidence of economic as well as clinical benefit in order for the product or service to be paid for.

National government formularies are also very strong buyers, and new products may face even higher "acceptance" hurdles than existing ones. As government budgets are strained (by, for instance, trade deficits—an example of interdependency with the macroeconomic dimension), price and utilization controls on the biomedical industry are sure to increase.

Substitutes. Substitutes for pharmaceuticals, for example, might include nutritional supplements, homeopathic remedies, herbal medicines, and so on. The ultimate substitute would be prevention of the disease or condition in the first place by, for example, lifestyle changes (encouraged by MCOs), or gene therapy. A substitute that must always be considered in the analyses is the option of *no consumption* of the biomedical product or service, for whatever reason (see also the discussion under "Sociocultural Dimension," p. 55).

Potential Entrants. Potential entrants are those organizations not currently operating as rivals but which may, nevertheless, discover and develop (1) competing products at lower cost, or (2) products providing additional benefit for the same cost. How vulnerable the biomedical firm is to such potential entrants depends, in this perspective, on the "size" of the barrier to entry that can be erected and/or maintained. All other things being equal, the higher the barrier, the greater the protection from entrants.

Barriers to Entry. All barriers are related to the cost of doing business, some more directly than others. One barrier directly related to cost is the *capital cost of entry*—the bricks and mortar, equipment and technology, as well as expert labor a biomedical firm requires simply to "open its doors." Discovering and developing therapeutic or diagnostic products require an enormous R&D outlay. In the pharmaceuticals sector, for instance, annual R&D budgets of $1 billion are common among the largest firms and constitute a sizable barrier.

Another barrier directly related to cost is the *absolute cost advantage* associated with owning, or having protected access to, important inputs, including labor (see "Suppliers," above).

Less directly related to cost, but a very important factor in the biomedical industry, is *regulation*. Medicines, diagnostics, and devices may be the most heavily regulated industrial products, and a firm's expertise in dealing with and meeting the requirements of regulatory bodies may constitute a competitive advantage. Especially with regard to clinical development, experienced and financially well-off pharmaceuticals companies may be in an advantageous position compared with inexperienced, less financially well-off firms. That is one explanation for the number of development alliances between biotechnology and pharmaceuticals companies.[8]

Patent protection, which provides essentially monopoly rights (and "rents") to the company for the duration of the patent (usually 20 years from the date of filing), constitutes a protective barrier. But, because even slightly different chemical formulations may be legally patentable, this barrier may be modest (unless the company files many and strong patent applications). Patents on genetically engineered substances, on the other hand, are often subject to protracted disputes. In any case, the effective market life of a patented product is about 10 years, because it may take 10 years from patent filing to approval for the market.

Access to distribution channels—to how the product reaches the end customer—can constitute a barrier if the biomedical firm is well-connected to physician opinion leaders, or to managers of large formularies, or to pharmacy benefits management firms, etc.

In general, the biomedical firm's ability to *differentiate* its output to consumers may provide a barrier to entrants. However, differentiation refers only to subjective qualities, such as the perceived reputation for quality, innovation, and so on. The ability to differentiate depends on the buyer and the product. In health care, the patient is almost always a *lay buyer,* without scientific or clinical expertise. Moreover, the health care good or service is, for the patient, an *experience good.* It is difficult if not impossible to know ahead of time whether one will be satisfied or not with, say, a prescription antifungal medicine. One must experience the medicine to know that it worked (or failed). Under these conditions—lay buyer and experience good—differentiation along subjective lines is possible.

Other buyers, medical personnel and insurers, are *professional,* with technical expertise in choosing among competing products. For them, the biomedical product is a *search good*: they can decide on the basis of price and clinical and pharmacoeconomic studies (i.e., without consuming it) that the antifungal will or will not be a useful and worthwhile addition to the formulary. Under these conditions—professional buyer and search good—products can only be distinguished by "objective" characteristics such as pharmacoeconomic benefit.[9]

Finally, a sizable barrier can be erected by the biomedical firm's membership in the relevant *knowledge network*. As reviewed under the macroeconomic dimension, companies in the biomedical industry depend upon universities and public research institutions to help advance the relevant sciences. Because biomedical products are essentially knowledge embodied in a device or a compound, being "firmly connected to the invisible network of knowledge is critically important."[10] Of course, the size of this barrier depends on the strategic acumen to fund the *right* outside research and to encourage the *right* researchers to work in the firm's areas of interest. If the number of relevant research institutions is limited, the strategic leadership should move quickly in establishing such connections. This is addressed in Chapter Eight.

Figure 12 summarizes Porter's competition framework.

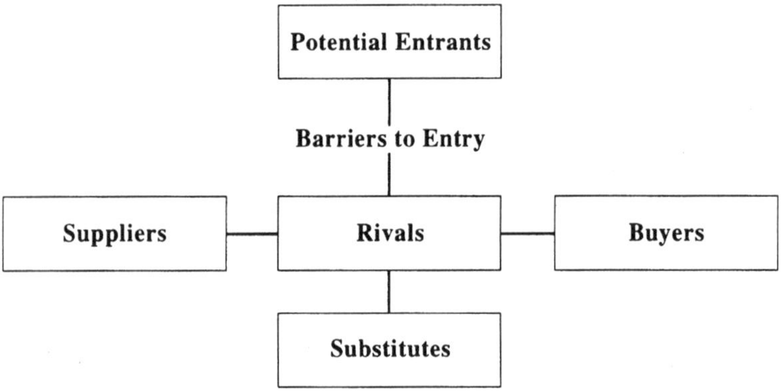

Figure 12. The competitive environment (Porter model).

Scientific and Technologic Dimension

KEY COMPONENTS: Science Frontiers, R&D and Infrastructure Technologies

Science Frontiers. In the mid-1970s, the biomedical industry was literally transformed by Boyer's and Cohen's recombinant DNA (rDNA) and Milstein's and Koehler's monoclonal antibody (MAb) techniques. Sciences relevant to the industry, which had advanced by what de Solla Price termed "scholarly bricklaying," became almost overnight a series of rapidly advancing science frontiers.[11] That is, insights into one problem, such as manipulating DNA, illuminated questions that were asked earlier but were still unanswered or only partially answered (e.g., can cells from two species be fused?). Knowledge from the new field of molecular biology cascaded throughout other disciplines, and new technologies emerged (e.g., biosensors). Advances followed one another so rapidly that one researcher commented to me: "Biology used to be like botany. You could read a textbook 50 years old and not much had changed. Today, texts on molecular biology can be outdated in 2 years."

One consequence of rapidly advancing science frontiers is the compression of product and process life cycles. The sciences underlying the biomedical product, the R&D technologies used to discover the product, and the process technologies used to make that product can become outdated quickly by new discoveries. (This is also addressed in Chapter Four.)

R&D and Infrastructure Technologies. Also transforming the biomedical industry have been R&D and infrastructure technologies. How work is accomplished is vastly different today from a decade earlier. Workstations that can model complex molecular structures and new combinatorial chemistry techniques allow drugs to be "designed" instead of randomly found by screening. Human receptors can be cloned as targets for new drugs. Libraries of molecular structures elucidated by NMR are available to scientists on networked PCs. The Internet allows members of "invisible colleges" to communicate daily if desired, rather than primarily at conferences. Miniaturization of electronic components and combinations of living and electronic components ("biosensors") also hold great potential for what is produced and how it is produced in the biomedical industry.

A rigorous and exhaustive review of these and other issues should also inform discussions about developing R&D resources, to ensure the biomedical firm's scientific and technologic advantage in the future. The discussion in Chapter Four is particularly relevant to analyses of this dimension.

Epidemiology

KEY COMPONENTS: **Current Diseases and Conditions, Emerging Diseases, Medical Need**

Analyses should include current diseases and conditions and their genetic components in the industrialized and nonindustrialized nations. Other components of this dimension are emerging diseases from resistant infectious organisms, incidence and prevalence of novel and rare diseases such as Ebola, and environmentally linked diseases or conditions. The purpose of such reviews is to balance the state of knowledge about disease with medical need, so that the diseases, conditions, mechanisms, etc., on which R&D should focus can be selected.

Demographics

KEY COMPONENTS: **Population Aging, Fertility, Ethnicity, Education, Income, Geographic Location**

Because of the collinearity of many diseases and conditions with age, epidemiology and population demographics are likely to be among the most familiar dimensions to people in the biomedical industry. But, the demographic dimension includes other components that are key, such as education, income, and geographic location. The latter three are often highly correlated with health status, with insurance status, with attitudes towards health care and medicines (discussed under "sociocultural"), and with ability to pay for health care goods and services. A rigorous and exhaustive analysis of all demographic factors should be conducted for every major national market in which the biomedical firm does or expects to do business.

Political–Regulatory Dimension

KEY COMPONENTS: **Strategic Technologies, Regulation, Price Controls, Utilization Controls**

Strategic Technologies. Today, *strategic* technologies are those deemed to have economic value as well as those deemed to have military value. Biotechnology, medical technology, and applied molecular biology have been identified as *strategic,* because they are

> . . . likely to have not only substantial economic impact but also very large infrastructure and social impact. . . . Furthermore, they each affect several indus-

try sectors and a multitude of products, processes, and services.[12]

As a result of this designation, the biomedical industry has been and will continue to be under pressure from government industrial policy (or science and technology policy) to *produce*. Especially since 1980, the response of major governments to strategic technologies has been very similar. They have identified public sector R&D likely to be the source of breakthrough innovations in order to ensure its support. And, they have designed incentives for private sector investment in, and development of, these innovations. A brief review of biotechnology will illustrate.

In 1981, Japan's Ministry of International Trade and Industry (MITI) selected biotechnology as one of the key technologies of the future. Under MITI's Agency for Industrial Science and Technology, three large-scale, multiyear, collaborative biotechnology research programs were funded. These programs were modeled on Japan's Very Large Scale Integration (VLSI) effort that had been so successful in capturing market share in microelectronics from U.S. and European firms.

Following shortly after Japan's announcement in 1981, the European Community formally identified biotechnology as one means of promoting global industrial competitiveness. A Community framework policy for investment in and encouragement of collaborative R&D in Europe (e.g., the Eureka program) was put in place in 1987.

In the context of this increasing attention, two U.S. House of Representatives committees requested an evaluation of America's commercial position in biotechnology. A 1984 report stated that three factors were crucial to maintaining leadership in this strategic technology: government support of basic and applied science, adequate numbers of trained personnel, and funds for industrial development.

Follow-on legislation to promote biotechnology included the Federal Technology Transfer Act of 1986, which provided incentives for Federal and industry scientists to participate in Cooperative Research and Development Agreements.

Interestingly, government funds are being used to promote public and private sector R&D in these strategic technologies. But the products—new medicines, devices, and diagnostics—are paid for by these same governments. As will be described below, governments and other health care insurers are putting increasingly stringent price and utilization controls on the biomedical industry.

Regulations. In addition to being strategic, the biomedical industry is one of the most highly regulated industries. Because these regulations are such an integral part of the business, they will not be addressed here. Suffice it to say that rigorous and exhaustive analyses would cover current and proposed regulations in every national market in which the biomedical firm does or expects to do business.

Price controls. Wherever governments pay for drugs, they exert some form of control on price. For the sake of balance, policies of several industrialized nations as well as the U.S. are briefly noted here.

In Japan, the Ministry of Health and Welfare publishes a table listing the official prices at which drugs will be reimbursed. Since the early 1980s, when pharmaceuticals accounted for nearly 40% of national health care expenditures, these prices have been continuously reduced. The cuts have also been selective. In order to make the domestic market less attractive to domestic firms (thus encouraging their globalization), the Ministry has targeted antibiotics, vitamins, and generics (areas of focus of domestic companies) for the largest cuts.

Germany legislates a reference price system, which sets the price at which insurance (the sickness funds) will pay for drugs. This system is intended to contain costs as well as to stimulate price competition among drug manufacturers. The Health Care Structure Reform Act of 1993 mandated nonnegotiable physician budgets for prescription pharmaceuticals (among others) for up to 3 years while reforms were designed and implemented.[13]

France regulates the margins of wholesalers and retailers, as a means of controlling drug prices, and has also imposed strict controls on a product-by-product basis. Compared with prescription drug prices in the U.S., France has nearly the lowest expenditure, second only to Spain (a difference of more than 100 points in the price index).[14]

The U.K. links manufacturer profit to the size of the firm's investment in Great Britain. Prices of individual drugs are not regulated, but once they are set based on the profit agreement, the government restricts drug manufacturers from increasing them. Government can also remove drugs from the list of reimbursable products, if prices are considered excessive.[15]

In Canada, the provincial governments have either frozen or reduced budgets for prescription drugs, imposed price freezes, and shifted the cost for drugs by imposing higher copayments on individuals (except for pensioners).

Despite their use for a number of years, the above pricing schemes have met with mixed success. A comparison of prescription drug price controls in Canada and Europe concluded that controls effectively kept individual drug price increases below overall inflation. However, such controls did not completely curtail the increase in drug spending.[16]

In the U.S., government currently exerts control over drug prices only for those purchased by the Department of Veterans Affairs (VA); those reimbursed by the federal/

state system of Medicaid, which pays for the health care of the indigent; and those reimbursed by Medicare, which pays for the health care of those over 65 years (and currently pays for inpatient drugs but only for outpatient drugs used in end-stage renal disease, such as erythropoietin). The VA requires manufacturers to sell at a sizable discount called the "depot price." Medicaid requires manufacturers to return an amount of money equal to the difference between the drug price as of October 1990 and the current average wholesaler's price, if the latter rose faster than inflation. Medicare, on the other hand, examines both the cost and effectiveness of drugs, diagnostics, and medical devices. The price of erythropoietin, for example, was set by Medicare after the government had completed an analysis of the internal cost data from the drug's manufacturer and determined a "fair" price.

Also exerting control over prices are managed care organizations (MCOs). Because they provide both health care and health insurance, MCOs have an incentive to manage expenditures by carefully managing care. Although manufacturers publish an average wholesalers price, the actual revenue the company receives depends on the distribution channel. Managed care channels account for nearly half the U.S. prescription drug volume and extract sizable discounts for that volume. In addition to managed hospital and retail pharmacies, managed care plans also make use of mail order pharmacy, especially for drugs for chronic illness. The volumes of drugs distributed through managed care channels, and the size of the discounts, are expected to increase.

Utilization controls. Controls on drug utilization include limited lists, common in Italy, Germany (where they are known as "negative lists"), the U.K., and Scandinavia, among others. Limited lists are, in effect, formularies of drugs that the government will reimburse. The latter nations also attempt to control utilization by influencing physician prescribing. In Germany, a sizable reduction in drug spending was believed to be attributable to (1) increased prescribing of less costly drugs, such as generics and drugs on the reference price system, and (2) reduced prescribing of drugs considered to be excessively or inappropriately prescribed. (Physicians, as noted above, were given a global budget for patient care, including pharmaceuticals.)

U.K. physicians are given data on their prescribing patterns, with benchmarks based on comparable average cost for the region, demographics, and taking into account inflation. They also receive information on the comparative prices and efficacy of drugs. Mandated generic substitution and strict formulary management are common utilization controls in Canada and the U.S.

In U.S. managed care plans, a group of physicians and pharmacists constitutes a formulary or technology assessment committee that regularly reviews each therapeutic class of medicines, examining both clinical outcomes data and spending on the major prescription drugs. For equivalent clinical quality among drugs, the decision of which to include in the formulary is made by the committee in favor of lower cost. It must be emphasized that these committees will tolerate some patient inconvenience and minor side effects, if the overall cost of care is reduced. MCOs also conduct physician profiling activities—examining, in other words, how each physician prescribes in comparison with other physicians—and will intervene if physicians prescribe nonformulary drugs excessively.

A control on both price and utilization that is expected to become widespread is disease management, which puts much of the financial risk of treating a particular disease on the biomedical firm.[17] Managed care plans will "carve out" one or more diseases and contract with a firm to provide drugs, devices, diagnostics, education, counseling, and so forth, at a per patient ("capitated") rate that has been negotiated by the plan. If the biomedical firm takes care of patients for less money than the negotiated rate, the difference can be kept as extra profit. But, if the cost of care exceeds the negotiated rate, the firm bears financial responsibility.

To summarize: Controls on price have generally reduced the rate of growth in price but not necessarily total spending. More effective have been utilization controls that attempt to encourage patients and physicians to consider the cost-effectiveness of alternatives. In all major industrialized nations, these are the controls to which the biomedical industry must respond.

Sociocultural Dimension

KEY COMPONENT: Attitudes

Deserving the same rigorous and exhaustive analyses as the components in preceding dimensions are people's attitudes. The reason for their importance is that attitude, by definition, represent a predisposition to behave in a certain way. Thus, a consumer's positive attitude towards herbal medicines (a *substitute*, in the microeconomic dimension) is likely to predispose him to purchase that product rather than a chemical or biological prescription drug. A consumer's negative attitude towards medical professionals is likely to predispose her to refuse a diagnostic procedure recommended by a physician. *Market size depends in part on people's attitudes in that market.*

A short list of attitudes to be explored should include attitudes towards:

- health
- sickness
- alternative therapies
- self-care
- medical professionals
- "big business" (e.g., biomedical companies)
- animal testing
- genetic engineering
- patient rights.

These should be analyzed for every market in which the biomedical firm does or expects to do business, and their possible consequences for research, clinical development, consumer marketing, and so on should be examined.

Figure 13 summarizes the seven dimensions and (some of) the key components.

PART II: ASSESSING TURBULENCE

Assessing levels of turbulence, dimension by dimension, is the crucial second part of this methodology. I have two reasons for my emphasis (in this chapter and throughout the book) on turbulence. First, turbulence implies uncertainty and imperfect information. The strategic leadership must focus people's attention on the most turbulent dimensions in determining, setting, and adjusting the coordinates of the R&D vectors. The reason is that the technology trajectory is more likely to be knocked off course—away from the strategic vision—in the most turbulent dimensions.

Second, in addition to presenting numerous hazards, the most turbulent dimensions also provide the raw materials for shaping the environment, for envisioning the future and creating it (instrumental technology strategy). When

Macroeconomic	Microeconomic	Scientific/ technologic	Epidemiologic	Demographic	Political-regulatory	Sociocultural
Investment, trade, national budget, currency, employment	Rivals, suppliers, buyers, substitutes, potential entrants, BTE	Science frontiers, R&D and infrastructure technologies	Current diseases and conditions, emerging diseases, medical need	Aging, fertility, ethnicity, education, income, location	Strategic technologies regulation, price and utilization controls	Attitudes

Figure 13. Dimensions of environment and some key components.

people have a prepared mind for the richness associated with a turbulent dimension, and consciously monitor the appropriate weak signals of change, they are better able to take advantage of opportunities and less likely to be caught off guard.

The level of turbulence is a composite estimate described by three parameters:

- *Rate of Change.* How slowly or quickly will the key components of the dimension change over one to two decades?
- *Magnitude of Change.* How similar or different will the future state of the dimension be in comparison with the present state, because of changes in the key components?
- *Unpredictability of Change.* How certain is it that these changes in the key components will occur?

Low levels of turbulence imply a slow rate of change, low magnitude, and low unpredictability. High levels of turbulence imply the contrary.

For each dimension, the issues and consequences identified in the key component analysis should be "played out" over a 10- to 20-year time horizon and answers to the above questions estimated (this is similar to scenario analysis or scenario-based planning). A matrix, such as that illustrated in Figure 14, is useful. In this matrix I have arbitrarily assigned values of 1 and 2 to "s(low)," 3 to 5 to "moderate," and 6 to 9 to "high," so that an index for each dimension (or subset of components, if the dimension is very complex) may be derived. The index, of course, has no intrinsic meaning but is a useful device to help people come to agreement on levels of turbulence. The spread in

	(S)low		Moderate			High			
	1	2	3	4	5	6	7	8	9
Rate of change									
Magnitude of change									
Unpredictability of change									

For each dimension, come to agreement on the levels of the following:

RATE: How slowly or quickly will the key components of the dimension change over one or two decades?

MAGNITUDE: How similar or different will the future state of the dimension be in comparison with the present state, because of changes in key components?

UNPREDICTABILITY: How certain are people that these changes in the key components will occur?

Figure 14. Assessing turbulence.

values from "moderate" to "high" emphasizes the strategic importance of high levels of turbulence.

Consider the microeconomic dimension and a hypothetical pharmaceuticals company as an example. After studying the issues and consequences identified under rivals, buyers, suppliers, potential entrants, and barriers to entry, the strategic leadership of this firm decide that the most important events for their particular company over the next two decades will be the following (the most important issues and events should differ from firm to firm):

- Consolidation results in a few super-rivals in the pharmaceuticals sector, making competition within therapeutic areas extremely intense.
- Pharmacoeconomic evidence is required for admission of drugs to formularies internationally, raising the cost of, and changing the skill mix for, clinical trials.
- New centers of excellence in the firm's core technologies emerge in Asia, especially China.

People completing the analyses agree that the *rate* of change in the microeconomic dimension will be moderate to high (they decide upon a value of 6), the *magnitude* of change high (9), and the *unpredictability* of change moderate (5), for an overall index of 20 (Figure 15). In contrast, for events in the demographic dimension over the same period, they decide that the rate of change would be slow (1), the magnitude moderate (3), and the unpredictability low (1), for an overall index of 5.

It must be emphasized that the level of turbulence is a social construct; it is an artifact of people's knowledge about specific issues at a given time. Judgments of turbulence must be reexamined and subject to expert review as often as necessary, to avoid over- or underestimation biases. The judgments and their underlying assumptions must also be made explicit and transparent to all in the firm.

Given peoples' best judgment about levels of turbulence, the strategic leadership in the example pharmaceuticals company would then establish for every dimension, but especially for the more turbulent dimensions, a set of *weak signals* to be closely monitored. For instance, they decide that weak signals relevant to the development of new centers of excellence in China might include (1) weakening of the U.S. dollar, (2) continued reduction in the rate of growth of U.S. funding of basic research, and (3) increasing numbers of

	(S)low		Moderate			High			
	1	2	3	4	5	6	7	8	9
Rate of change						X			
Magnitude of change									X
Unpredictability of change					X				

Figure 15. Turbulence of the microeconomic dimension (pharmaceuticals example).

European and U.S. post-docs going to China to pursue their studies.

The purpose of establishing a set of weak signals, confirmed by experts in the respective dimension, is to prepare people to take steps when those signals reach an agreed threshold level. In the example firm, the strategic leadership decide to invest $5 million in funding several research alliances with major Chinese universities immediately. They will increase that investment to $25 million, when the number of U.S. and European educated post-docs reaches 20 in a particular university. Their purpose is to stay well-connected to the relevant knowledge network and ensure a strong competitive position.

As the above scenario implies, the cost of responding to a *weak* signal is low, and the benefit is that the biomedical firm can gain competitive advantage. As signal strength increases, so, too, does the cost of responding. By the time an

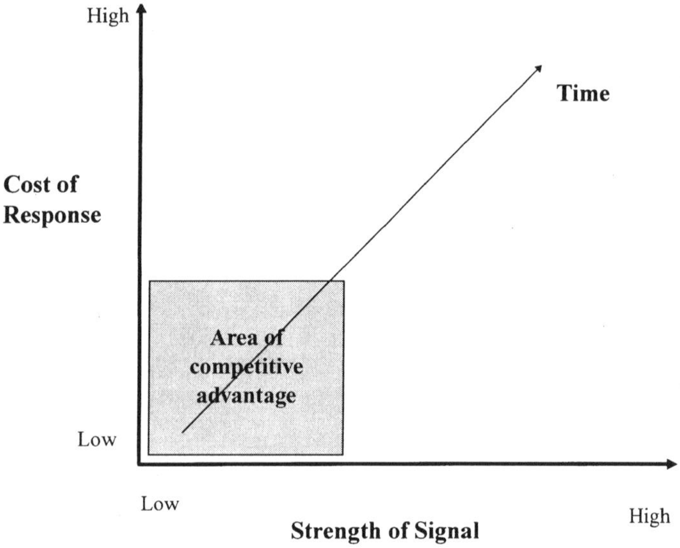

Figure 16. Cost of response to signals.

issue is obvious to everyone in the industry, the cost of a strategic response could be prohibitive (Figure 16).

PART III: FINAL ASSESSMENT

Interdependence

The first part of the methodology involves analyzing the dimensions of environment ("space"), rigorously, exhaustively, and *separately*, by identifying key components and the likely changes in those components over a 10- to 20-year period. The second part involves an assessment of the levels of turbulence as if each dimension were a discrete force acting on the biomedical firm, and establishment of a set of weak signals to monitor.

The dimensions are, of course, interdependent. Events in the macroeconomic dimension have an impact on price and utilization controls on the biomedical industry (political–regulatory dimension), on consumers' income (demographic dimension), and on publicly funded research (scientific and technologic dimension). People's attitudes (sociocultural dimension) have an impact on market size and competition (microeconomic dimension). Reduced federal budgets (macroeconomic dimension) have an impact on the response of government buyers of biomedical products and services (microeconomic dimension).

The concluding part of my methodology consists of two parts. In the first, the degree of interdependence among dimensions at the time of the analysis are determined. Using an interdependency matrix such as that shown on Figure 17 for the hypothetical pharmaceuticals company, the estimated degree of interdependence between pairs of dimensions is assigned a rank of low, moderate, or high. As I said earlier, examining interdependence is likely to reveal that a dimension initially viewed as stable may actually be turbulent, because of its interdependence with more turbulent dimensions. The objective of this first step is to help the strategic leadership decide if a reranking of dimensions is warranted, again to focus attention.

In the hypothetical pharmaceuticals company, people decide that events in the macroeconomic dimension have a large impact on price and utilization controls on the biomedical industry (political–regulatory). They judge the degree of interdependence between these dimensions to be high and continue with similar discussions until the matrix is completed. As a result of their deliberations, they agree that the demographic dimension's *moderate* degree of inter-

	Macroeconomic	Microeconomic	Sci/Tech	Epidemiological	Demographics	Political–Regulatory	Sociocultural
Macroeconomic							
Microeconomic	Low to moderate						
Sci/Tech	Moderate	Low					
Epidemiological	Low to moderate	Low	Low to moderate				
Demographics	Moderate	Low	Low to moderate	High			
Political–Regulatory	High	Moderate	Moderate to high	Low	Moderate		
Sociocultural	Low	Low	Low to moderate	Low to moderate	High	Moderate	

Events in the macroeconomic dimension impact price and utilization controls on the biomedical industry (political–regulatory) and consumer's income (demographic). Thus; the degree of interdependence between macroeconomic and political–regulatory dimensions is rated *high*; between macroeconomic and demogragphic dimensions, *moderate*. The degree of interdependence between the demographic and epidemiologic dimensions, as well as between demographic and sociocultural, is rated *high*, and so on.

Figure 17. Interdependency matrix (pharmaceuticals example).

dependence with the macroeconomic dimension (e.g., unemployment and health insurance) and *high* degree of interdependence with epidemiology make it less stable than their initial index suggested. (Remember, the group assigned the demographic dimension a turbulence index of "5.")

Another benefit to thinking about interdependence in this way, the group discovers, is finding additional weak signals to monitor. They decide that in- and out-migration patterns resulting from geopolitical instability is an important weak signal applicable to the demographic and epidemiologic dimensions.

Impact and Unpredictability

The second step in the final assessment of space–time is for the strategic leadership to select the most important issues, trends, and events surfaced by *all* prior discussions and come to shared agreement as to the impact of these issues and events on the firm (low or high) and their perceived unpredictability (low or high). Figure 18 illustrates four generic technology strategies appropriate to each of the four combinations of low or high impact and unpredictability. The purpose of this second step is to ready the biomedical firm for action at a time when the cost of responding is low and the probability of preempting competition is still high (cf. Figure 16).

For instance, the strategic leadership of the hypothetical pharmaceuticals company decide that, given their historic strengths in antiinfective R&D, the increasing resistance of organisms is an issue of low impact and low unpredictability. The corresponding technology strategy, *business as usual* (Cell 1), requires everyone in R&D to monitor relevant

		Impact	
		Low	**High**
Unpredictability	**Low**	Business as usual 1	2 Maintain competitive position
	High	Monitor weak signals carefully 4	3 Gain competitive advantage

Figure 18. Generic technology stategies.

weak signals for clues that the levels of impact and unpredictability might be changing. Resistant organisms might, for example, alter immune system characteristics in a manner that could have high impact on future discovery targets.

Although issues and events may have low impact and low unpredictability at one point in time, no one should assume they will remain inconsequential in the future.

The leadership decide that a trend of high impact and low unpredictability is increasing enrollment of populations in both Europe and the U.S. in some type of managed care plan. The corresponding technology strategy, *maintain competitive position* (Cell 2), requires the leadership to ensure that the benefits of their firm's products outweigh the costs. They believe that new information technologies are needed immediately, to improve the efficiency and lower the cost of clinical trial data collection and analyses.

An event of high impact and high unpredictability is the emergence of centers of scientific excellence in China. The corresponding technology strategy, *gain competitive advantage* (Cell 3), requires the leadership to invest in research alliances with Chinese universities while signal strength is low (cf. Figure 16). As signal strength increases, they will adjust their investment accordingly.

Finally, the strategic leadership decide that an event of low impact and high unpredictability is resurgence in popularity of Soviet-style governance and redrawing of Soviet national boundaries. The corresponding technology strategy, *monitor weak signals carefully* (Cell 4), requires them to be ready to increase or decrease the firm's clinical research presence in these geographic areas, depending on the change in weak signals.

CONCLUSION

Analyses of the space through which the biomedical firm will be propelled over time must continue as long as the firm endures. After the strategic vision is defined and the technology identity is selected, the "endless loop" of the instrumental strategic process consists of the three-part methodology described above: taking experimental actions (such as the pilot programs noted earlier), evaluating feedback from these actions, taking action based on feedback, etc. Strategic resources—intellect, time, money—must be devoted to these analyses.

If everyone has a mind prepared for the richness associated with a turbulent environment, the biomedical firm will be better able to gain competitive advantage, and the technology trajectory will be less likely to be knocked off its strategic course.

NOTES

1. Kenneth Andrews used the term, "dimension," to describe a part of the firm's environment (*The Concept of Corporate Strategy*, Dow-Jones-Irwin, Homewood, IL, 1980).
2. The concept of turbulence was applied to the social (as opposed to physical) sciences by F. E. Emery and E. L. Trist (e.g., *Towards a Social Ecology*, Plenum Press, New York, 1973). The concept was later applied to the environment of modern organizations by H. Igor Ansoff (*Implanting Strategic Management*, Prentice-Hall International, Englewood Cliffs, NJ, 1984).
3. Scenario analysis such as described in this chapter has been called a "powerful technique" for preparing a firm for the future (H. Leufkens, "Scenario Analysis: A Powerful Technique to Identify Appropriate R&D Strategies," Centre for Medicines report, 1995, U.K.). See also H. Leufkens, F. Haaijier-Ruskamp, A. Bakker, and G. Dukes, "Scenario Analysis of the Future of Medicines," *Br. Med. J. 309* (1994), 1137–1140.
4. Organisation for Economic Cooperation and Development, *OECD Economic Surveys: United States*, OECD Publications, Paris, p. 50.
5. Organisation for Economic Cooperation and Development, *Science and Technology Policy: Review and Outlook 1994*, OECD Publications, Paris, 1994.
6. National Institutes of Health, *Data Book 1994*, U.S. Department of Health and Human Services, Bethesda, MD, 1995.
7. Michael Porter, *Competitive Strategy*, The Free Press, New York, 1980.
8. Ernst & Young, *Biotech 95: Reform, Restructure, Renewal*, Ernst & Young, Palo Alto, CA, 1995.
9. Differentiation is discussed in general microeconomic theory by Paul A. Samuelson and William D. Nordhaus, *Microeconomics*, McGraw-Hill, New York, 1992. The more detailed analysis of lay and professional buyers, and search and experience goods, is based on Douglas F. Greer, *Industrial Organization and Public Policy*, Macmillan, New York, 1980.
10. The "knowledge network" was described with reference to the pharmaceuticals industry in Europe by M. L. Burstall and I. Senior, in *The Community's Pharmaceutical Industry*, Office for Official Publications of the European Community, Luxembourg, 1985.
11. Derek L. deSolla Price, in *Little Science, Big Science . . . and Beyond*, Columbia University Press, New York, 1986. Price also talked about

"invisible colleges" joining scientific researchers (discussed under "R&D and Infrastructure Technologies," p. 47).

12. This designation can be found in a number of sources, including the U.S. Department of Commerce report, *Emerging Technologies*, Washington, D.C., 1990 (p. 7).
13. U.S. General Accounting Office, *German Health Reforms*, Washington, D.C., 1994.
14. U.S. General Accounting Office, *Prescription Drugs: Spending Controls in Four European Countries*, Washington, D.C., 1994.
15. U.S. General Accounting Office, *Prescription Drugs: Companies Typically Charge More in the United States than in the United Kingdom*, Washington, D.C., 1994.
16. U.S. General Accounting Office, *Prescription Drugs: Prices and Regulations in Canada and Europe*, Washington, D.C., 1994.
17. K. Lumsdon, "Hard labor: Disease management," *Hospitals & Health Networks 5* (April 1995), 34–42.

CHAPTER FOUR

ABOUT TECHNOLOGY

The purpose of this chapter is to lay the groundwork for R&D vector decisions that will result in effective use of existing science and technology resources to achieve business objectives and development of those resources to ensure future competitive advantage. The chapter defines technology and the characteristics of core technologies, describes three properties associated with technology, and illustrates how the strategic leadership can utilize an understanding of these properties to map R&D investments to the technology life cycle. By appropriate mapping, the traps of overinvesting in mature, well-known technologies and underinvesting in emerging ones will be avoided.

PROLOGUE

Achieving the biomedical firm's current business objectives, maintaining competitive position, and gaining future competitive advantage—e.g., from high-impact, high-unpredictability events (Figure 18)—depend upon intelligent allocation of science and technology resources. In turn, intelligent allocation of people, time, facilities, equipment, and money means that:

- the investment in mature, well-known technologies is appropriately low, and
- the investment in emerging technologies is appropriately high.

To put this another way, the current and future success of the biomedical firm depends in large measure on two conditions. First, the strategic leadership are able to map R&D investments to the technology life cycle (Figure 19). Second, those investments are consistent with the strategic vi-

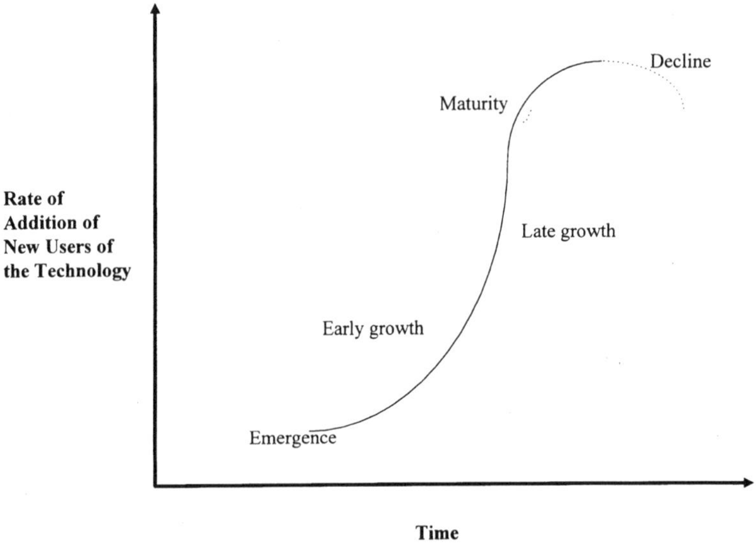

Figure 19. Technology life cycle.

sion and chosen technology identity. I address the first condition in this chapter; I address the second in Chapters Five, Six, and Seven.

Charting a course through turbulent space–time demands an understanding of technology at a level of sophistication consistent with instrumental strategy. Without that sophisticated understanding ("rocket science"), as the example of Solar Imaging in Chapter Two illustrated, there are two traps that can jeopardize the biomedical firm's competitive success. In Solar Imaging, R&D managers in Division X *overinvested in mature* technologies, while those in Division Y *underinvested in emerging ones*. Some of the consequences of their inability to map R&D investments to the technology life cycle appropriately included:

- loss of key scientists from R&D
- loss of market share for the company as a whole
- failure to achieve key corporate objectives
- designation of the company as an attractive acquisition candidate.

This chapter lays the groundwork for competitive technology strategy, the blueprint for decisions on determining, setting, and adjusting coordinates of the R&D vectors. Competitive technology strategy results in effective *use* of the biomedical firm's existing science and technology resources to achieve business objectives and *development* (advancement) of those resources to ensure future competitive advantage.

In the following section, *technology* and *core technologies* are defined. Then, I discuss three properties associated with technology and their likely combinations. In the concluding section, I illustrate how the strategic leadership can utilize an understanding of these combinations to map

R&D investments to the technology life cycle appropriately. By appropriate mapping, the traps of overinvesting in mature technologies and underinvesting in emerging ones will be avoided.

SOME DEFINITIONS

A sophisticated understanding of technology must begin with an appreciation of both the distinctions and the relationships among *knowledge, science,* and *technology.*

Knowledge refers to that which individuals and, because of them, relevant parts of the R&D organization, understand. Knowledge can be explicit (e.g., the structure of DNA) or tacit (e.g., gene sequencing techniques that R&D staff have acquired by experience and do "by feel"). Tacit knowledge is implicit understanding and is extremely important, because it can be a powerful competitive advantage to the biomedical firm. However, because it is tacit, it requires "on the hoof or 'person embodied' mechanisms of transfer."[1] In other words, R&D staff have to work directly with the individual(s) who possesses tacit knowledge, in order to acquire it. Tacit knowledge, by definition, cannot be transmitted via written instructions. By the same token, tacit knowledge can "walk out" of the firm, if the people who possess it decide to leave.

Science refers to codified or systematized knowledge and is, by that definition, explicit (in other words, science is always *communicable*).

Technology, discussed in more detail below, refers to science-in-use. Technology, like science, is always explicit. If something cannot be communicated, it remains tacit knowledge.[2]

There is no necessarily linear progression *from* knowledge *to* science *to* technology but rather a circuitous, iterative, or sometimes step-function progression.[3] Technology can lead directly to advances in knowledge (e.g., X-ray crystallography in Watson and Crick's work) as readily as knowledge can lead directly to advances in technology (e.g., production of monoclonal antibodies or MAbs).

During this period of rapidly advancing science frontiers (the scientific and technologic dimension, Chapter Three), biomedical technologies "are strongly science-related . . . [and] are now particularly intimate activities" with both knowledge and science.[4] That is why the strategic leadership must be as concerned with generating new knowledge as they are with advancing science and technology.[5]

Staying connected to the relevant knowledge networks can maintain a high barrier to potential entrants and is a critical means of ensuring the biomedical firm's future competitive advantage.

Below, I discuss *technology* and the *core technologies* of R&D in more detail.

Technology

At the most rudimentary level, technology has been defined as "the *means* utilized to accomplish a task, . . . [including] *performance programs* stored in individuals' memories."[6] Technology can be said to possess both a hardware aspect (i.e., an equipment or tool component) and a software aspect (i.e., knowledge of the equipment's or tool's use).[7]

The range of R&D technologies relevant to the biomedical firm depends on its *domain*. This encompasses the products and services, populations, and means by which products and services are and will be produced.[8] Domain, in other words, is specified by those decisions guiding the allocation of business resources within the framework of corporate strategy (Chapter Two).

Metaphorically speaking, the biomedical firm's domain is embedded in a technological network. Every organization can be said to utilize, in effect, a subset of technologies from the "chains or threads of technology" relevant to that industry.[9]

I have three reasons for discussing both domain and technological network in a chapter on technology: First, *domain* is a matter of strategic choice and can be changed. The strategic leadership could decide not to continue producing generic drugs; or to enter the prescription drug business from a cancer diagnostic capability. Such decisions change the biomedical firm's domain.

Second, because the domain is embedded in a technological network, every strategic decision to change domain also changes the biomedical firm's "place" in that network. In other words, a change in domain exposes the organization to different technological possibilities—a consequence not always appreciated.

Third, in this model of strategy, the technological network provides a vital part of the raw materials to be shaped in the achievement of the strategic vision (the scientific and technologic dimension, Chapter Three). The firm's technology trajectory both passes through this network over time and is affected by it, as it is by every other dimension of that space. Being instrumental as far as technology strategy is concerned requires, among other things, that the strategic leadership see beyond *what is* in the subset of technologies that is utilized to *what might be* in that subset. Creative decisions about R&D technologies—drawing, in other words, from outside the traditional subset of technologies—are hallmarks of innovating organizations.[10]

Core Technologies

Although every biomedical firm makes use of multiple technologies from the technological network, not all of these are *core technologies*. Core technologies have been defined as those that should be most protected from external influences. They should be kept and/or developed in-house as much as possible, so as to reduce "the influence of the environment on the technological core."[11]

The term, *core technology*, should be distinguished from that of *core competence*. A core competence can include, but is not limited to, a technology. Competence of the biomedical firm encompasses culture and values, as well as knowledge and expertise.[12] A core technology, science-in-use, is more narrowly defined as only biomedical knowledge and expertise.

My focus in this book is on the core technologies of R&D. Such technologies will have the following characteristics:

- *They represent a sizable amount of tacit knowledge within the biomedical firm.* There are no externally published instructions that have yet captured what R&D scientists are capable of accomplishing.
- *Their "software" is more important than their "hardware."* There may be little in the way of tools or equipment. But, even if there is a sizable component of equipment, its use has not yet been mastered outside the biomedical firm.
- *They are very close to ("intimate" with) the respective leading-edge knowledge and science frontiers.* It is likely that R&D scientists are working intently with public sector researchers or researchers outside the biomedical industry.

Core technologies, like aspects of the domain, are a matter of strategic choice. Examples could include assay as well as information technologies, combinatorial technologies as well as clinical trial design. Whatever the technologies and wherever—discovery or development—they are employed, it is critical that they be identified. Outsourcing or sharing them in joint ventures, etc., will introduce significant contingencies and expose the biomedical firm to unnecessary external influence. As a general rule, outsourcing or sharing must be avoided when the technology is *core*. The reason for this statement will become clearer in the following discussion.

THREE PROPERTIES

The next step in developing a sophisticated understanding of technology, always with the objective of making the right resource allocation decisions in R&D, is for the strategic leadership to become familiar with three properties. Every technology can be described by particular levels of *uncertainty* and *equivocality*, and the contingent property of *risk*. These are not properties of technology *per se* but rather properties associated with a technology *in the minds of those who use it* (remember: technology is science-in-use). Each property is discussed below.

Uncertainty

If all elements of a technology were perfectly known, there would be no uncertainty. Uncertainty means that some knowledge about the technology is lacking; information, in other words, is imperfect. The level of uncertainty, in turn, depends upon two conditions at a given point in time.

First, how well are the relevant *raw materials* (both physical materials and data) understood? Second, how explicit is the *process* by which raw materials are *converted* to output?

The higher the level of uncertainty, "the greater the amount of information that has to be processed by decision makers during execution of the task."[13] When the level of uncertainty is high, the task itself will be "ill defined, . . . [there will be] problems in understanding the raw materials, . . . [and it will not be] possible to develop wholly explicit operations to convert" raw materials into output.[14]

High levels of uncertainty are predictably characteristic of a technology during the emerging and growth stages of the life cycle, when the knowledge and science bases are being expanded.

For a technology in these stages, a number of weak signals will apply:

- Information about the technology will reside primarily within a few scientific circles (the "invisible colleges" noted in Chapter Three). If published, information will be restricted to one or a few journals. Communication about the technology will be mostly word-of-mouth among specialists in the emerging field and concerned with the numerous questions remaining about the raw materials and/or conversion processes.
- If applicable, there will be few issued patents.
- Experience with the technology will be limited to relatively few organizations, primarily academic and government.
- Cost data will be scarce, and they will vary widely.

As illustration, consider the case of biotechnology. In 1973, rDNA techniques were first used successfully by U.S. academic scientists; in 1975, researchers in a U.K. government laboratory produced MAbs. Although there was tremendous excitement in scientific circles, information about the techniques themselves, the raw materials, and the conversion process was scarce. One of the critical problems in the diffusion of biotechnology in Europe was described as limited information on genetic engineering.

Few patents existed; in fact, there was considerable debate over whether genetically engineered materials could be patented. Experience with the technology was initially confined to a few major academic and government laboratories. Costs of developing biotechnology products, and the revenues that might be generated, varied so widely that one reporter declared: "[Biotechnology has] cast off its ropes to the solid deck of science and drifted into the warm southern seas of publicity."[15]

According to these weak signals, biotechnology was characterized by high levels of uncertainty throughout most of the 1970s. An example of a current technology with a high level of uncertainty might be the fusion of biological organisms and microelectronic devices to create biosensors.

Equivocality

Uncertainty is concerned primarily with the *how to* aspects of a technology; that is, with the raw materials and processes for converting them into finished outputs. Equivocality is concerned primarily with the *for what* aspects.

Equivocality is a term used by communication researchers to describe the existence of multiple and conflicting interpretations about some phenomenon.[16] With regard to a technology, equivocality refers to multiple and conflicting interpretations about its *use, role,* and *scope.*

When the level of technology uncertainty is high, more information about raw materials and conversion processes must be sought and exchanged, in order to reduce uncertainty to manageable proportions. But, when the level of technology equivocality is high, "it is not clear what questions to ask, what kinds of information are needed, or even what information sources will help to bring about a workable level of clarity."[17] Reducing equivocality requires face-to-face communication among those concerned, until shared agreement is reached.

Studies of the U.S. Microelectronics and Computer Technology Corporation (MCC) showed that, when interpretations of a technology varied among users, that technology was very difficult to transfer from the MCC to an organization that would develop it further. The solution was bringing scientists and other stakeholders together, in order to facilitate discussions that achieved some common ground.[18] In other words, face-to-face meetings reduced equivocality.

High levels of equivocality are predictably characteristic of a technology during the emerging and growth stages of the life cycle, when experience with applications of the technology are still limited.

The predominant weak signal indicating high equivocality is multiple and divergent opinions, in scientific and trade publications and relevant knowledge networks, about the use, role, and scope of the technology.

It is important to understand that, for some period during the growth stage of the technology life cycle, the level of uncertainty may be lower than the level of equivocality—or, the level of equivocality may be lower than the level of uncertainty. During the emerging stage, levels of both properties will be high. But, the level of one is likely to be reduced faster than the other. If raw materials become better understood and conversion processes better specified more quickly than opinions about the use, role, and scope of the technology converge, the level of uncertainty associated with that technology will be lower than the level of equivocality. Conversely, there may be general agreement about the use, role, and scope of a technology at the same time that major questions about the raw materials and/or conversion processes remain unanswered. In this case, the level of equivocality associated with that technology will be lower than the level of uncertainty.

For instance, by the early 1980s, U.S. doctoral students in molecular biology were routinely performing rDNA experiments. However, there were still multiple and conflicting interpretations of the use, role, and scope of the technology in the biomedical industry. Some believed that biotechnology represented a rich, new discovery tool; others, that biotechnology had only limited utility. According to the criteria described thus far, at that time the level of uncertainty was lower than the level of equivocality.

From a life cycle perspective, reduction of either uncertainty or equivocality indicates the point of inflection between the *early* and *late* growth stage. But, it is worth considering that the time required to reduce equivocality may be less than that required to reduce uncertainty. If uncertainty is reduced first, there may be a shorter span remaining in that technology's growth stage. If equivocality is reduced first, that span may be longer. (Reaching agreement about a technology may take less time than gaining knowledge about it.)

Risk

The level of this third property is contingent on the levels of the other two. Risk is defined as how *near* or *far* the output of a technology is to use by a subsequent function or by the end consumer.[19] A proxy measure for this conceptual distance is time: the further an output is from some consumer (including a subsequent function within the firm), the longer the time required to produce that output and the higher the probability of failure to do so.

Because it is a contingent property, high levels of risk are predictably characteristic of a technology during the emerging and into the late growth stages of the life cycle, when uncertainty or equivocality or both are high as well.

The level of uncertainty is high when people do not yet understand the *how to* of the technology. The level of equivocality is high when people do not yet agree on the *for what*. In both instances, it will be some time (greater "distance") before the output from the relevant technology will be usable as input by a subsequent function or end consumer. High levels of uncertainty and/or equivocality are associated with a high probability of technological failure. Thus, the weak signals either for high uncertainty or for high equivocality are the weak signals for high risk.

In the example of biotechnology, although the uncertainty of the basic techniques had declined by the early 1980s, the level of risk was still high because of the high equivocality. Only a few genetically engineered products had moved from discovery to development (and fewer still to the end consumer). Biotechnology company stock volatility reflected that high level of risk.

It must be emphasized that levels of uncertainty, equivocality, and risk are social constructs. As such:

- The levels depend on people's understanding of, and experience with, a technology at a given point in time.
- Different organizations can accumulate understanding of and experience with a technology at different rates, depending on (among other things) their strate-

gic vision, technology identity, and allocation of resources.

If the strategic leadership have made an explicit commitment to defining the leading edge of relevant sciences and technologies (the technology identity), then R&D resources must be allocated so that the biomedical firm's scientists accumulate understanding at a faster rate than competitors' scientists.

It is the potential for reducing the levels of technology uncertainty and/or equivocality (and, by implication, risk) faster or slower than competitors that enables the biomedical firm to gain, or lose, competitive advantage.

It should now be clearer that *core technologies* will be, by definition, in the emerging and growth stages of their life cycle. By the time a technology is mature, it can no longer be considered *core*. Core competence, on the other hand—such as industrial fermentation or purification techniques—may make use of mature technologies, and the caveat about outsourcing does not apply.

In the concluding section, I describe the likely combinations of the three properties. Understanding what combinations are logical and to what stage of the life cycle they apply allows the strategic leadership to map R&D investments appropriately. In this way the traps of overinvesting in mature technologies and underinvesting in emerging ones (e.g., Solar Imaging) will be avoided.

MAPPING R&D INVESTMENTS TO THE LIFE CYCLE

As I stated above, the levels of uncertainty and equivocality associated with a technology will be highest during the emerging stage of its life cycle, but one property may be reduced faster than the other during the growth stage. For some technologies, uncertainty is reduced faster than equivocality (e.g., biotechnology). For others, equivocality is reduced faster than uncertainty (e.g., biosensors). In all cases, the level of risk will be contingent on which level, uncertainty or equivocality, is higher.

Practically speaking, the important levels are *high* and *low*. For the purpose of allocating R&D resources, the strategic leadership should be most concerned about whether the level of uncertainty or equivocality is high (indicating emerging technologies) or low (indicating mature technologies). With an understanding of the three properties and appropriate scanning of weak signals, they should also be able to discern the points of inflection on the life cycle indicating that a technology is likely to enter the next stage.

When any technology is about to enter the mature stage, new technologies in the emerging stage of their life cycle may be evident (Figure 20). Assuming a lively connection

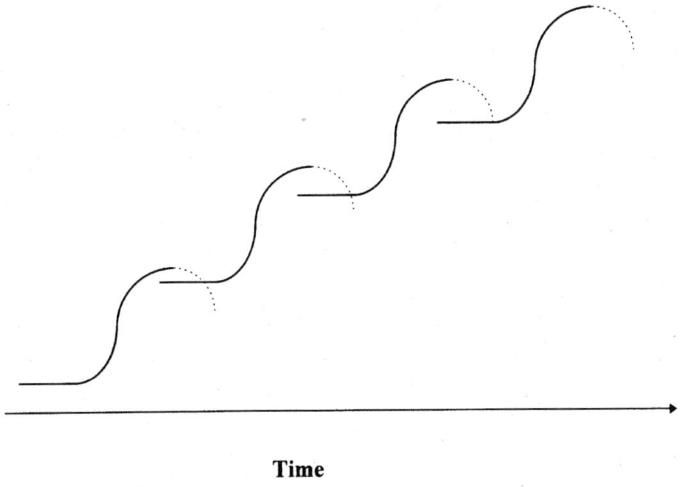

Figure 20. Evolution of technologies.

to the knowledge networks, scientists in R&D should be aware of the weak signals indicating that another, newer (potential core) technology could replace the one that is maturing and no longer to be considered core. As a rule of thumb, the strategic leadership must begin investing in selected future core technologies *before* the mature stage of the "old" core technology.

Given high and low levels of the two independent properties, four combinations are possible (Table 1), and each combination can predictably be associated with a life cycle stage (Figure 21). Mapping R&D investments in a particular technology to its life cycle must be based on peoples' best judgment as to the levels of uncertainty and equivocality associated with that technology in the larger technological network (i.e., among those who now use it).

If the decision is made to reduce the level of "technology uncertainty" within the biomedical firm faster than it might be reduced by the competition, the strategic leadership must allocate sufficient resources to helping R&D scientists gain a better understanding of the raw materials and/or conversion processes.

If the decision is made to reduce the level of "technology equivocality" within the firm faster than it might be reduced by the competition, the strategic leadership must al-

Table 1. Combinations of Properties Associated with Technology

Possible Combination	Uncertainty	Equivocality
1 (risk high)	High	High
2 (risk high)	High	Low
3 (risk high)	Low	High
4 (risk low)	Low	Low

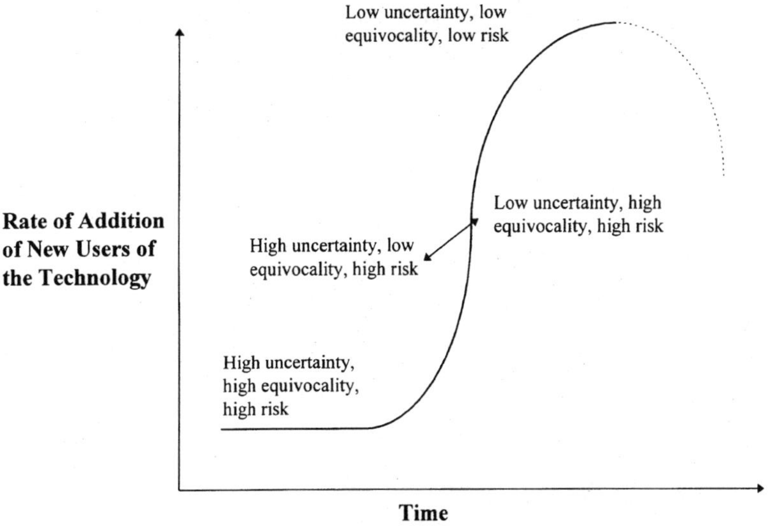

Figure 21. Three properties and the technology life cycle.

locate sufficient resources to *bringing people together* within R&D, until reasonable agreement is reached as to the use, role, and scope of the technology. It is extremely important that the strategic leadership understand and appreciate the different activities that must be supported, to reduce these two properties in such a way as to produce competitive advantage.

Below, I discuss each combination and the general recommended investment decisions.

High Uncertainty, High Equivocality

When the levels of both uncertainty and equivocality (and, thus, risk) are high, a technology is in the emerging stage of its life cycle. At this time, the strategic leadership have the *largest* window of opportunity to be instrumental at the lowest cost. To help R&D scientists gain experience with and understanding of that technology, resources should be allocated to pilot projects in-house and/or in collaboration with public sector researchers. The objective should be to reduce the levels of both uncertainty and equivocality within the biomedical firm faster than within actual or potential competitors. The strategic leadership may also consider acquiring an organization that has already developed the desired capability (this is discussed in Chapter Eight). Note that reducing the level of only uncertainty or only equivocality will give the biomedical firm a tenuous competitive advantage, because the level of risk (probability of failure) will remain high.

High Uncertainty, Low Equivocality

When the levels of uncertainty and risk are high but equivocality is low, a technology is in the growth stage of its life cycle, around the point of inflection between early and late growth. Although understanding of and experience with the technology will still be limited, a low level of equivocality implies that shared agreement about the use, role, and scope of the technology has already been reached. The window of opportunity for instrumental investment will be *smaller* than in the first scenario.

The strategic leadership must allocate resources to ensure that scientists within R&D reduce the level of uncertainty (and risk) faster than competitors do. The activities supported must be such that people gain a better understanding of the raw materials and/or conversion processes involved. The cost of reducing uncertainty at this stage of the life cycle will be higher than in the above example. As I stated in Chapter Three, when signal strength rises, so does the cost of responding. At this time the firm is, in effect, close to the edges of the area of competitive advantage shown in Figure 16.

Low Uncertainty, High Equivocality

When the level of uncertainty is low but the levels of equivocality and risk are high, a technology is in the growth stage of its life cycle, again at the point of inflection between early and late growth. However, the strategic leadership now have the *smallest* window of opportunity for instrumental investment, because only a lack of shared agreement as to the use, role, and scope of the technology keeps the level of risk high. As I noted above, the time required to reduce the level of equivocality may be less than that required to reduce the level of uncertainty. Thus, if the level of uncertainty is reduced first, there will be a shorter span of time during which the technology will remain in the growth stage.

In this case, the strategic leadership must ensure that scientists within R&D are technically proficient and then allocate resources so that face-to-face communication occurs efficiently among those who must come to shared agreement as to the use, role, and scope of the technology. The firm is now, in effect, on the border between competitive advantage and disadvantage (Figure 16), so astute decisions are critical.

Low Uncertainty, Low Equivocality

When levels of both uncertainty and equivocality are low, the technology will be in the mature stage of its life cycle and can no longer be considered *core*. It is now too late for instrumental investment in this technology; other firms have already begun to invest in new technologies in their emerging and growth stages. The challenge for the strategic leadership now is to avoid overinvestment in a "posthumous technology."[20]

TECHNOLOGY IDENTITY AND TECHNOLOGY INVESTMENT

I stated in Chapter Two that the strategic vision for the firm must, in the new model of strategy, include technology identity. The latter I defined as one of three general orientations towards science and technology that will characterize research and development efforts: defining the leading edge, remaining on the leading edge, or pursuing the leading edge. Each identity, in turn, governs resource allocation and can be restated as a generic R&D vector decision.

First, the strategic leadership can choose to allocate the firm's science and technology resources so that levels of uncertainty and equivocality (and risk) associated with the technology within R&D are *reduced faster* than they are within competitors. This choice is consistent with a technology identity of *defining the leading edge.* People within the biomedical firm attempt to gain competitive advantage by building up knowledge and expertise more quickly than competitors are capable of building them. The stage of the technology life cycle at which the cost of this decision is lowest is the emerging stage.

Second, the strategic leadership can choose to allocate the firm's science and technology resources so that levels of uncertainty and equivocality (and risk) associated with the technology within R&D are *consonant with* the levels in competitors' organizations. This choice is consistent with a technology identity of *remaining on the leading edge.* Competitive position is maintained by building up knowledge and expertise within the biomedical organization at about the same rate as competitors build them.

Third, the strategic leadership can choose to allocate the firm's science and technology resources so that levels of uncertainty and equivocality (and risk) associated with the technology within R&D *lag* the levels in competitors' organizations. This choice is consistent with a technology identity of *pursuing the leading edge.* Although this choice is appropriate under certain circumstances (discussed in Chapter Six), it is *never* appropriate for core technologies.

The so-called preparatory section of the book ends here. The following chapters stand on this foundation and address the "how-to" of competitive technology strategy.

NOTES

1. A very useful overview of knowledge, science, and technology is provided by W. Faulkner and J. Senker in *Knowledge Frontiers,* Oxford University Press, Oxford, England, 1995. The quotation is from p. 38. In their book they also review England's experience with "public sector research and industrial innovation in biotechnology, engineering ceramics, and parallel computing."
2. Like Faulkner and Senker, above, D. Leonard-Barton discusses differences among knowledge, science, and technology, particularly with regard to communication ("The Intraorganizational Environment," in F. Williams and D. V. Gibson (Eds.), *Technology Transfer: A Communication Perspective,* Sage Publications Inc., Newbury Park, CA, 1990).
3. Faulkner and Senker refer to de Solla Price's "two stream" model, whereby new science generally builds upon prior science and new technology upon prior technology. Historically, there is very little evidence of a linear progression *from* knowledge *to* science *to* technology (D. de Solla Price, *Little Science, Big Science . . . and Beyond,* Columbia University Press, New York, 1986).

4. This is drawn from Faulkner and Senker's study of the biotechnology industry (see Note 1, above), p. 29.
5. Two recent authors who emphasize the importance of knowledge generation are D. Leonard-Barton (*Wellsprings of Knowledge*, Harvard Business School Press, Boston, MA, 1995) and I. Nonaka, and H. Takeuchi (*The Knowledge-Creating Company*, Oxford University Press, New York, 1995).
6. This fundamental definition is from D. Gerwin, "Relationships Between Structure and Technology," in P. S. Nystrom and W. Starbuck (Eds.), *Handbook of Organizational Design*, Oxford University Press, Oxford, England, 1981.
7. The distinction between a technology's "hardware" and "software" comes from E. Rogers, *Diffusion of Innovation*, The Free Press, New York, 1983.
8. One of the most seminal writers with regard to technology and organizations studies is J. D. Thompson who, among other terms, defined organizational domain (*Organizations in Action*, McGraw-Hill, New York, 1967).
9. This comes from a very interesting study of British firms by K. Clarke, D. Ford, and M. Swan, "Company Technology Strategy," *R&D Management 19* (1989), 215–239. The quotation is from p. 225.
10. This was also noted by Nonaka and Takeuchi, Note 5, above.
11. The definition of core technologies is taken from Thompson, Note 8, above, p. 22.
12. A useful discussion of core competence is provided by R. A. Goodman and M. W. Lawless, in *Technology and Strategy*, Oxford University Press, New York, 1994.
13. The work of J. Galbraith in the 1970s was enormously important to the study of properties associated with technology (*Designing Complex Organizations*, Addison-Wesley, Reading, MA, 1973). The quotation is from p. 71.
14. D. Gerwin (Note 6, above, p. 16) expanded the definition of uncertainty, drawing also on the work of C. Perrow ("A Framework for the Comparative Analysis of Organizations," *American Sociological Review 32*(2) (1967), 194–210).
15. This book captures all the drama of the first Initial Public Offerings of U.S. biotechnology companies (R. Teitelman, *Gene Dreams (Wall Street, Academia, and the Rise of Biotechnology)*, Basic Books, New York, 1989). The quotation is from p. 40.
16. Originally discussed in the context of "media richness," the notion of equivocality has passed into much more general currency. One

seminal work on equivocality was by R. Daft and R. Lengel, "Information Richness: A New Approach to Managerial Behavior and Organization Design," in B. Staw and L. Cummings (Eds.), *Research in Organizational Behavior*, JAI Press, Greenwich, CT, 1984.

17. My colleague and I considered the impacts of uncertainty and equivocality in project communication (D. Stork and A. Sapienza, "Task and Human Messages Over the R&D Project Life Cycle," *Journal of Project Management XXIII*(4) (1992)). The quotation is from p. 67.
18. This is an interesting example of the application of "equivocality" to, in this case, technology transfer (C. M. Avery, and R. W. Smilor, "Research Consortia," in F. Williams and D. V. Gibson (Eds.), *Technology Transfer: A Communication Perspective*, Sage Publications, Inc., Newbury Park, CA, 1990).
19. The notion of "technology distance," or time until some output is ready, comes from Avery and Smilor's work, Note 18, above.
20. My friend and colleague, J. Desmond Fitzgerald, used this phrase ("Reflections on Some Problems in the Management of Drug Discovery," in F. Gross (Ed.), *Decision Making in Drug Research*, Raven Press, New York, 1983).

CHAPTER FIVE

DETERMINING COMPETITIVE POSITION

The purpose of this chapter is to present a framework for assessing both the internal consistency of the firm's science and technology investments and the competitive advantage these investments may create. The chapter reviews basic conditions expected to be met and skills expected to be in place by this point in the strategic process, presents a methodology for codifying R&D investments and determining the competitive position of the firm from the perspective of biomedical science and technology, and introduces three technology strategy archetypes derived from that codifying methodology. Subsequent chapters build on the methodology and archetypes.

STOCK TAKING

The preceding chapters about strategy, environment, and technology laid the foundation for creating competitive advantage. By this point in the strategic process, I assume that the following basic conditions are met within the biomedical firm:

- acceptance of and appreciation for the necessity of being instrumental (i.e., attempting to shape the future), the model of strategic trajectories, and technology strategy as the blueprint for determining, setting, and continuously adjusting R&D vector coordinates
- shared agreement on and adherence to the strategic vision
- commitment to the values implied by that vision (and to a plan that addresses any gap between desired values and actual behaviors)
- shared agreement on the chosen technology identity of R&D and the crucial role of programs and projects ("booster rockets")
- acceptance of and appreciation for the external environment as both dynamic forces acting on the technology trajectory and the source of raw materials to be shaped by instrumental technology strategy
- shared agreement on the selected organizational domain
- sophisticated understanding of the properties of uncertainty, equivocality, and risk and the weak signals by which their levels may be assessed.

I also assume that the corresponding skills and abilities are in place within R&D. In other words, the strategic leadership are able to:

- analyze key components of the multidimensional space through which the technology trajectory will pass
- identify the most turbulent dimensions at any given

time, the corresponding weak signals that must be monitored, and the area of competitive advantage for the firm (Figure 16)

- assess interdependence among dimensions
- select the issues and events of most importance to the firm and categorize them according to their impact and unpredictability
- implement the appropriate generic technology strategy, given the above categorization (Figure 18)
- assess the technological possibilities implied by the selected organizational domain
- distinguish between core competencies and core technologies
- determine life cycle stages of R&D technologies, in order to map investment appropriately.

The next step in the process, the subject of this chapter, is to determine the competitive position of the firm from the perspective of biomedical science and technology. Although a vital step, it is particularly difficult to accomplish. One study of British firms found that few managers were "able to assess their [organization's] technological assets and strengths," or to understand "the nature of [the organization's] technology position and the technological network of which their firm forms a part."[1] I believe that the reasons for this difficulty are, first, an incomplete understanding of the nature of technology (which I addressed in Chapter Four) and, second, the lack of a framework and metrics for evaluating and comparing technological assets and strengths.

As part of the "how to" of competitive technology strategy, I provide such a framework and metrics in this chapter. Below, I describe a methodology that uses vector quantities to help the strategic leadership evaluate the internal consistency of R&D investments and determine the competitive position of the firm from the perspective of biomedical science and technology. Then I illustrate the application of vector quantities, using a series of hypothetical firms in the medical device and pharmaceuticals sectors. In the concluding section, I introduce three technology strategy archetypes derived from the codifying methodology and discuss their implications for both internal consistency and competitive position. Subsequent chapters build on the methodology and archetypes.

VECTOR QUANTITIES

As I described in Chapter One, every discovery effort, program, and project can be depicted by a vector, and the sum of all efforts is the R&D vector. Over time, the R&D vectors describe a series of points that, when joined, define the technology trajectory (Figure 3). Consistent with the model of strategic trajectories, the metrics I propose for codifying science and technology investments are the *vector quantities* of origin, magnitude, and direction.

Origin

The starting point for evaluating and comparing the biomedical firm's science and technology efforts and capabilities, the origin, is the monetary amount allocated to a particular effort. If overall R&D efforts are being analyzed, the origin is set at the firm's total annual R&D budget. Competitors will include organizations with similar annual budgets that can then be compared with regard to the vector quantities of magnitude and direction (discussed below). If therapeutic, target, or product area efforts are being analyzed (e.g., programs and projects within one therapeutic area), the origin is set to that amount. Competitors will include organizations making a similar investment, *regardless of their overall R&D budget*. Setting the origin in this manner means that every analysis—of overall R&D efforts or of a therapeutic, target, or product area effort—corresponds to an investment plane within three-dimensional space (Figure 22).

It should be clear that the origin represents the tangible resources (facilities, equipment, expert labor, etc.) that are the basis for competitive advantage. As I noted in Chapter Three (under "Microeconomic Dimension"), tangible science and technology resources constitute a potential barri-

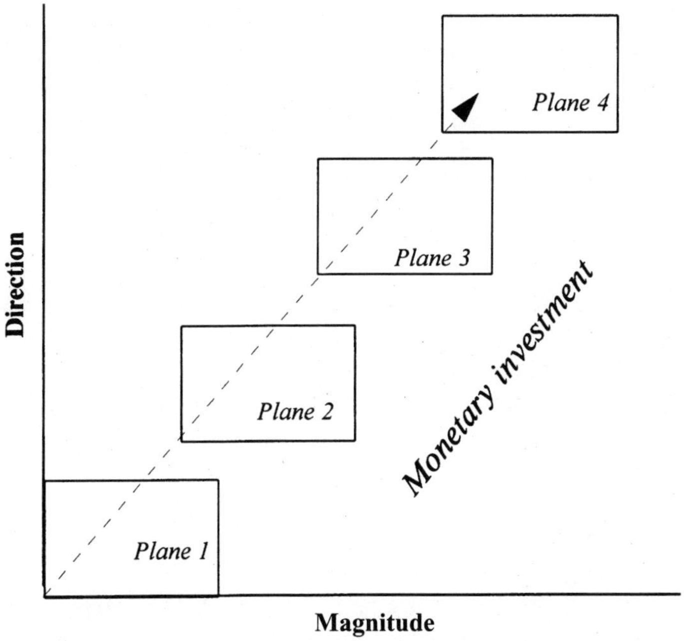

Figure 22. Investment planes.

er to the entry of competing organizations. But that barrier is only potential. It is the intangible *experience resource*—how, in other words, the tangible investment is used—that creates real competitive advantage.

Investing in tangible science and technology resources is a necessary but insufficient condition for competitive advantage. The leadership must also invest in building "strategically relevant skills and capabilities."[2] To the extent that they invest in *in-house knowledge, skills, and capabilities,* they help to create the experience resource. Experience in this sense is the superior ability "of the organization to learn, adapt, change, and renew over time" by the process of "search, problem finding, and problem solving" in R&D.[3] The experience resource is particularly relevant to core technologies (Chapter Four). In this framework, the experience resource of the biomedical firm is represented by the vector quantities of magnitude and direction.

Magnitude

Magnitude reflects the size of the internal capacity associated with a particular effort. This vector quantity is estimated by the proportion of the relevant investment (origin) devoted to in-house ("make") versus external ("buy") capacity. The make-or-buy choice affects the biomedical firm's long-term competitive advantage under turbulent conditions. A predominantly "buy" choice in R&D will cost less initially and may be used to provide limited exposure to new technologies. On the other hand, more resources allocated to external organizations mean fewer are available to develop in-house expertise. And returns to the firm may be reduced by royalty and other fees.

Licensing-in a sizable proportion of internal capacity is a

short-term tactic at best and definitely *not* appropriate for core technologies. The strategic leadership must understand the properties associated with technology decisions (Chapter Four), in order to make timely investments in internal knowledge, skills, and capabilities during the emerging and growth stages of selected future core technologies. There is a limited window of investment opportunity, beyond which the money will be wasted. Chapters Six and Seven also address this topic.

Direction

The third vector quantity, direction, reflects the risk profile of the particular effort and is estimated by the proportion of the relevant investment (origin amount) devoted to scientifically and technologically risky efforts. (Remember from Chapter Four that risk is contingent on the levels of uncertainty and/or equivocality.) Magnitude and direction together estimate the depth of people's experience (i.e., internal capacity) and the risk willingness of the particular portfolio of activities.

It is important to point out that the risk profile of the overall R&D portfolio (as distinct from therapeutic, target, or product area portfolio) should be determined by, and should remain consistent with, the selected technology identity:

- defining the leading edge
- remaining on the leading edge
- pursuing the leading edge of relevant sciences and technologies.

Technology identity sets the context for determining what proportion of science and technology resources should be allocated to high-risk efforts (i.e., efforts to define the leading edge), moderate-risk efforts (to remain on the leading edge), and low-risk efforts (to pursue the leading edge). Certainly, the leadership must balance investments so that survival of the biomedical firm is not jeopardized. Not all resources can be allocated only to high-risk or only to low-risk efforts for an extended period of time. On the other hand, a larger proportion of resources must be invested in high-risk efforts if the technology identity is to define the leading edge than would be required if the identity were to pursue the leading edge.

There must be internal consistency among (1) strategic vision; (2) technology identity; (3) proportion of science and technology resources allocated to internal knowledge, skills, and capabilities; and (4) proportion of science and technology resources allocated to each risk category in the relevant portfolio. Vector quantities, applied below, provide a useful strategic tool for assessing this internal consistency and for determining the competitive position of the firm from the perspective of biomedical science and technology. More detailed allocation guidelines are given in Chapter Six.

VECTOR QUANTITIES APPLIED

In this section, I provide simple examples of the application of vector quantities to overall R&D efforts as well as to a particular therapeutic area effort. The examples are for the purpose of illustrating how these quantities reveal a biomedical firm's competitive position. They are not meant to do justice to the time and energy that must be spent on data gathering, analyses, and (especially) discussion within the firm. (The examples do presume that a sophisticated strategic intelligence capability exists within the organization.)

First Illustration—Medical Devices

The first application of vector quantities is to four hypothetical medical device firms, including Solar Imaging from Chapter Two. All firms ("L," "M," "Solar Imaging," and "P") have invested about $75 million in R&D this year; they have annual sales of about $1 billion. In the medical device sector, the largest firms have annual sales of about $5 billion and R&D investments of about $350 million. Small firms have sales in the hundreds of millions and invest about 7% of that in R&D.[4] With regard to tangible science and technology resources, these hypothetical midsize firms have an advantage primarily over the small companies. This is not meant to imply that firms investing much more will necessarily be more creative, or that one of the hypothetical firms cannot be catapulted into the big league because of technological innovation and sales success.

Because they are investing about the same amount in overall R&D, these firms' vectors will share the same origin (i.e., all four will be on the same investment plane). Industry strategic intelligence from public sources (including the Internet) as well as proprietary sources such as targeted analysts' and in-house scientists' trip reports suggests that the proportion of total R&D invested in internal ("make") capacity by these firms is as shown in Table 2.

The four companies specialize in diagnostic imaging devices and have fairly similar competing products. Within that specialty, estimates of technological risk converge on the following:[4]

Table 2. Proportion of Medical Device R&D Invested in Internal Capacity

Firm	% In-house ("Make")
L	40
M	56
Solar Imaging	19
P	21

- high risk: telemetry linking of devices to centralized data bases, with decision support systems
- moderate risk: nuclear medicine and ultrasound devices, no telemetry
- low risk: X-ray devices, no telemetry.

Based on this categorization, strategic intelligence suggests that the proportions of overall R&D invested in high-risk efforts are as shown in Table 3. The R&D vectors are listed in Table 4 and are plotted in Figure 23. (In this and all other examples, vector length is calculated as the square root of X^2 [i.e., in-house capacity] + Y^2 [i.e., risk profile].)

Evaluation. The first objective of those engaged in such analyses is to assess the internal consistency of their own

Table 3. Proportion of Medical Device R&D Invested in High-Risk Efforts

Firm	% High Risk
L	22
M	61
Solar Imaging	30
P	57

Table 4. Medical Device R&D Vectors

Firm	Vector Length
L	46
M	83
Solar Imaging	36
P	61

science and technology investments. Below, a very brief scenario is presented for each hypothetical medical device firm, with summary comments on internal consistency. A more detailed discussion of internal consistency is provided in Chapter Six.

In Firm L, approximately 40% of total R&D has been invested in in-house capacity and 22% in high-risk efforts. The strategic vision for this organization includes being among the top five firms, by sales, in every major national diagnostic imaging market. The chosen technology identity is to *remain* on the leading edge of the relevant sciences and technologies, and the strategic leadership believe that this identity is consistent with and supports the achievement of their vision. According to the above analyses, however, they question whether a sufficient proportion of R&D is allocated to high-risk efforts. They believe that the proportion allocated to in-house capacity could also be increased. The decision of how much to shift investment will depend both on their technology identity and on the firm's competitive position, addressed in the next subsection.

In Firm M, 56% of overall R&D has been invested in in-house capacity and 61% in high-risk efforts. The strategic vision for this organization includes entering the big league within 5 years. This requires "movement" to a different investment plane, because the largest firms invest around $350 million currently in R&D, compared with this

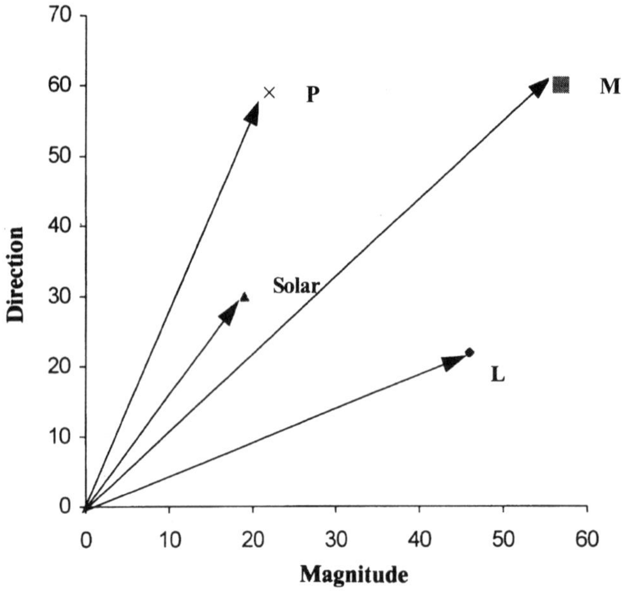

Figure 23. Medical device R&D vectors.

firm's $75 million. The strategic leadership have decided to divest noncore businesses to fund medical device R&D expansion. The chosen technology identity is to *remain* on the leading edge of the relevant sciences and technologies. According to the above analyses, the leadership agree that investing heavily in high-risk efforts is appropriate, given their desired change in competitive position. The current competitive position of Firm M is discussed in the next subsection.

In Solar Imaging, only 19% of overall R&D has been invested in in-house capacity and 30% in high-risk efforts. As might be inferred from the discussion of the firm in Chapter Two, there is not yet a shared strategic vision, although there is general agreement on ambitious sales growth targets over the next 5 years. Nor is there an explicit choice of technology identity. The strategic leadership have embarked on an aggressive in-licensing campaign, predominantly in what the industry would characterize as low- and moderate-risk efforts. There is no discussion of internal consistency of investment (or its lack). Their competitive position is described in the next subsection.

In Firm P, 21% of overall R&D has been invested in in-house capacity and 57% in high-risk efforts. The strategic vision for this organization includes being among the top 10 firms globally. The strategic leadership have decided to change the firm's technology identity from remaining on to *defining* the leading edge of the relevant sciences and technologies. Based on the above analyses, they are concerned about the possible inconsistency between what they have invested in in-house capacity (21% of R&D) and what they have invested in high-risk efforts (57% of R&D). They believe that their licensing strategy is "in the right direction" but that they should develop more expertise in-house. Their objective is to build capacity in selected future core technologies that will support the new technology identity. Their competitive position is discussed below.

Competitive Position. At the time of these analyses, Firm M appears to be in the strongest competitive position of the group for two reasons. First, there is internal consistency among:

- shared vision—entering the big league of the medical device industry by 2005
- technology identity—remaining on the leading edge of relevant sciences and technologies
- internal skills and capabilities—more than half of total R&D is invested in in-house expertise.

More than half of total R&D is allocated to high-risk efforts, in accord with their decision to enter the big league. Second, the company's R&D vector (Figure 23) is larger than that of the other firms.

On the other hand, the strategic leadership intend to move the firm from the current investment plane to that of the big league. Thus, their competitors-to-be are the largest firms in this sector. In addition to monitoring the activities of current competitors, the strategic leadership must conduct similar analyses of those top firms at least on an annual basis, to ensure that the intended technology trajectory of Firm M is "aimed" at the new endpoint (strategic vision) in the new investment plane. More detail on how this firm might change its competitive position is given in Chapter Seven.

Solar Imaging appears to be in the weakest competitive position within the group, and not only on the basis of the length of their R&D vector (Figure 23). In addition to the likely competitive disadvantage of their experience resource, there is not yet awareness within the strategic leadership of the need for internal consistency. If the leaaership do not examine their investments with more sophistication than heretofore, Solar Imaging could become an acquisition candidate (perhaps for Firm M). I discuss steps that the company might take to remedy this situation in Chapter Six.

Firms L and P illustrate interesting contrasts in competitive position. Firm L has the lowest proportion of R&D invested in high-risk efforts (22%) and could "fall back" to an effective technology identity of *pursuing* the leading edge of relevant sciences and technologies. Based on their analysis, the strategic leadership within Firm L decide to shift the allocation of resources so that at least one-third of overall R&D is invested in high-risk efforts within 3 years. They believe that, until the imbalance between depth of in-house capacity (i.e., 40% of R&D) and risk willingness of the portfolio (i.e., 20% in high-risk efforts) is corrected, Firm L is competitively vulnerable.

In Firm P, by contrast, a larger proportion of overall R&D is invested in high-risk efforts (57%) but only 21% is invested in internal capacity. The strategic leadership want to change the technology identity from *remaining on* to *defining* the leading edge of relevant sciences and technologies but have been relying on in-licensing to accomplish this. When they realize the inadvisability of this tactic for core technologies (Chapter Four), they are faced with the necessity of increasing the depth of internal knowledge, skills, and capabilities. They also question whether the proportion invested in high-risk efforts is sufficient, given the new technology identity (this is addressed in the concluding section).

Second Illustration—Pharmaceuticals

The second application of vector quantities is to six hypothetical pharmaceutical firms, including Lucida NeuroSciences from Chapter Two ("Lucida," "B," "C," "D," "E," and "F"). I begin by analyzing these firms' overall R&D efforts; next, I analyze a particular therapeutic area.

Overall Science and Technology Resources. All six hypothetical firms are midsize, with current annual R&D budgets of about $250 million and sales of about $1 billion. The largest firms in this sector have annual R&D budgets of more than $1 billion and sales of about $10 billion. Small firms have sales in the hundreds of millions and invest about 10% of sales in R&D.[5] Like the medical device example, the six pharmaceutical companies share the same investment plane.

Industry strategic intelligence from public and proprietary sources suggests that the proportion of total R&D invested in internal capacity by these six companies is as shown on Table 5.

Unlike medical devices, which can be used to diagnose and treat a number of diseases and conditions, medicines are produced for essentially "water-tight" therapeutic areas and diseases and conditions within these areas. In the pharmaceuticals sector, risk profile can be generally estimated by the proportion of resources allocated to various therapeutic areas. According to the current state of knowledge about the etiology and mechanisms of major diseases (i.e., scientific uncertainty), therapeutic area research can be ranked from higher to lower risk as follows: central nervous system diseases (CNS), cancer (CA), cardiovascular diseases (CV), and infectious diseases (AI, "antiinfective"). Within each area, of course, specific diseases and conditions may be ranked differently (e.g., AIDS research would be judged of higher scientific uncertainty than infectious disease research overall).

Based on this categorization, industry strategic intelligence suggests that the proportions of overall R&D invested in high-risk therapeutic areas are as shown on Table 6.

Table 5. Proportion of Pharmaceuticals R&D Invested in Internal Capacity

Company	% In-house ("Make")
Lucida	52
B	17
C	38
D	30
E	24
F	40

Table 6. Proportion of Pharmaceuticals R&D Invested in High-Risk Efforts

Company	% High-Risk
Lucida	48
B	27
C	35
D	53
E	20
F	18

The overall R&D vectors at this point in time are listed in Table 7 and are plotted in Figure 24.

At the time of these analyses, Lucida appears to be in the strongest competitive position. The six rivals can be divided into two groups, based on the ranking of their R&D vectors. Lucida, Company D, and Company C belong in the first group, with vectors of 71, 61, and 52, respectively. Companies F, B, and E belong in the second group, with vectors of 44, 32, and 31, respectively. The strategic leadership in Lucida decide to monitor firms D and C carefully, particularly D. Note that Company D has a similar proportion of overall R&D invested in high-risk efforts (53% versus Lucida's 48%). If their leadership were to increase the

Table 7. Pharmaceuticals R&D Vectors

Company	Vector Length
Lucida	71
B	32
C	52
D	61
E	31
F	44

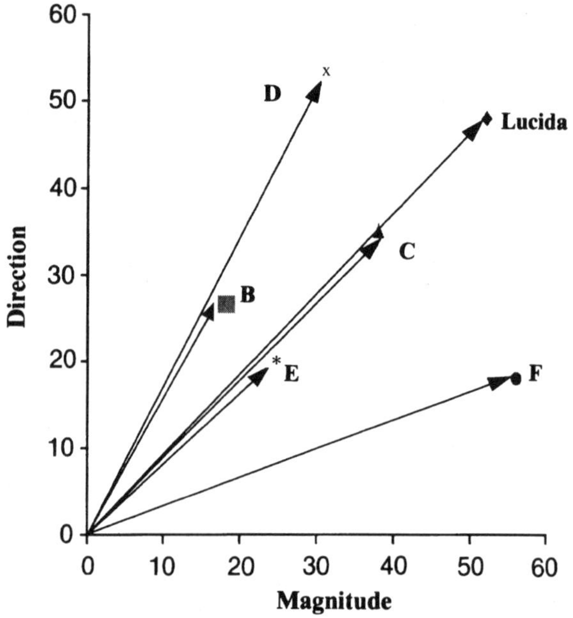

Figure 24. Pharmaceutical R&D vectors.

proportion of total R&D invested in internal capacity (beyond the current 30%), that firm could become a formidable competitor.

The strategic leadership in Lucida believe that Company C is also important to monitor, because its R&D vector is so closely aligned to that of Lucida (Figure 24). If the leadership in Company C were to increase investment both in high-risk efforts and in internal capacity, that firm could also become a strong competitor.

The remaining firms are deemed to be the weakest of the competitors at this time. Of course, that does not mean they can be ignored. In terms of vigilance, however, the strategic leadership in Lucida agree that Company D clearly warrants the greatest share. They decide to monitor this firm's R&D activity on a semiannual basis.

Because of the strategic importance of central nervous system research to Lucida (remember from Chapter Two that the leadership have decided on a niche strategy with an emphasis on central nervous system therapies), I continue the analyses of competitive position by evaluating that therapeutic area.

Therapeutic Area Resources. About 2 years ago, the leadership decided to focus on central nervous system therapies, with continuing efforts in cardiovascular and antiinfectives. As I described in Chapter Two, the strategic vision for Lucida is to be always among the top three firms in the selected niche, and their technology identity is to *define* the leading edge. A major business objective is to introduce a breakthrough therapy to the CNS market within 8 years. Currently, the amounts invested in research and in development in the three therapeutic areas are listed in Table 8.

For the sake of brevity in this chapter, I will analyze only CNS research. The strategic leadership will, of course,

Table 8. Lucida NeuroSciences R&D Investments

Research	$75 million
Central nervous system (CNS)	$25 million
Cardiovascular (CV)	$30 million
Antiinfective (AI)	$20 million
Development	$175 million
CNS (Phase III)	$95 million
CV (Phase II)	$10 million
AI (Phase III)	$70 million

complete analyses of research and development investments in each of the three therapeutic areas.

Of the rivals analyzed earlier, strategic intelligence suggests that Company D is investing a similar amount in CNS research. In addition, there are three biotechnology companies that must be included in this particular investment plane. The proportion of CNS research allocated to internal capacity by the five firms ("Lucida," "Company D," "Biotech 1," "Biotech 2," "Biotech 3") is as shown in Table 9.

Within a single therapeutic, target, or product area, of course, finer-grained data are needed to estimate the risk profile of that portfolio. For example, within CNS research, the strategic leadership must examine how funds are allocated to discovery programs *by disease or condition.* For this purpose, the leadership review a study of U.S. and Japanese pharmaceutical industry scientists.[6] As shown in Table 10, with close agreement, scientists had ranked 10 diseases and conditions highest in scientific uncertainty on a scale of 1 = lowest to 7 = highest. Highest scientific uncertainty was defined by an outside expert panel as consisting of:

- no well-developed animal models of the disease
- no chemical leads
- no validated preclinical test with high predictive

Table 9. CNS Research Investment in Internal Capacity

Company	% In-house Effort ("Make")
Lucida	70
Biotech 1	90
Biotech 2	60
Biotech 3	30
Company D	75

Table 10. Scientific Uncertainty of Pharmaceuticals Research

	Uncertainty (Mean Response)	
	Japan	U.S.
Neurodegeneration	5.504	5.886
Cognitive impairment	5.386	6.167
Autism	5.355	5.867
AIDS	5.315	4.727
Bone cancer	5.046	5.462
Lung cancer	4.969	5.231
Spinal cord injury	4.746	5.000
Hearing disorders	4.725	5.333
Neurologic injury	4.722	5.200
Hepatic encephalopathy	4.634	4.920

power of therapeutic utility in humans with the disease

- inability to identify dosage range using a biochemical marker
- ill-characterized pathophysiology of the disease ("unknown mechanism")
- ill-defined disease target.

Based on this categorization, Lucida's strategic intelligence suggests that the proportions of CNS research allocated to neurodegeneration, cognitive impairment, and/or autism discovery efforts are as shown on Table 11. The vectors corresponding to CNS research efforts are listed in Table 12 and are plotted in Figure 25.

At the time of these analyses, the strategic leadership in Lucida believe that their company is in a stronger competitive position than Biotech 2 and Biotech 3 but in a weaker position than Company D and much weaker than Biotech 1. According to several recent trip reports by their scientists, the $25 million or so that are invested in high-risk dis-

Table 11. CNS Research Invested in High-Risk Discovery

Company	% High-Risk
Lucida	40
Biotech 1	70
Biotech 2	23
Biotech 3	27
Company D	63

Table 12. CNS Research R&D Vectors

Company	Vector Length
Lucida	81
Biotech 1	114
Biotech 2	64
Biotech 3	40
Company D	98

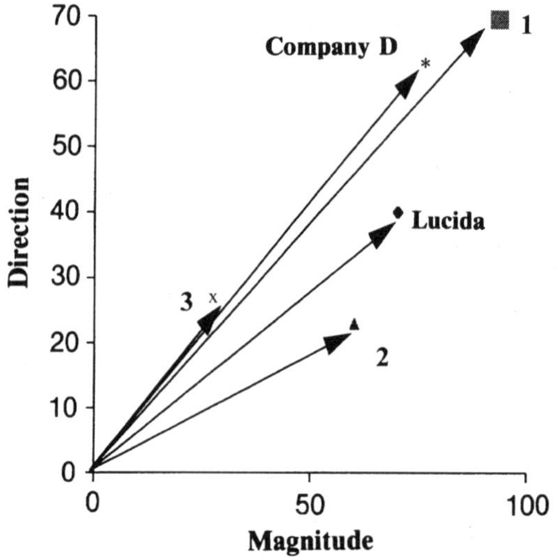

Figure 25. CNS therapeutic area vectors.

covery in Biotech 1 may represent more than 80% of that company's total research budget. The biotechnology firm has the largest experience advantage according to these measures, but most of its "eggs" are in this "basket."

In Company D, on the other hand, the $25 million is believed to represent about one-quarter of total research. This company may be less vulnerable than Biotech 1, because of the diversification of their research efforts. On the other hand, scientists in Biotech 1 may be able to preempt the competition. Given the volatile situation among the three closest competitors, the leadership in Lucida decide that this is a very turbulent dimension (Chapter Three—the microeconomic dimension). They set in place a sophisticated process for surveillance of these companies and review of the intelligence on a quarterly basis.

Summary. The strategic leadership in Lucida appreciate that an organization of their size must focus on fewer therapeutic areas than companies in the major league. But, based on the above analyses, they question whether their investments are really consistent with the strategic vision and chosen technology identity. As the discussion of technology strategy archetypes in the concluding section will suggest, their uneasiness is justified.

Issues

Before introducing the technology strategy archetypes, and their implications for internal consistency and competitive position, I want to acknowledge that the above applications of vector metrics are meant to raise a number of issues. Two major issues are addressed in subsequent chapters: internal consistency in Chapter Six, and changing competitive position in Chapter Seven. In this chapter, I want to address the issues of data, frequency of analyses, and number of competitors.

Data. Although the examples use other than round numbers, they are not intended to demonstrate precision. Vector metrics are part of a strategic *heuristic*. The objective is for the leadership to gather intelligent estimates of other firms' technology investments as well as their own. Data should be as "hard" as possible without "analysis paralysis."

Frequency of Analyses. As might be expected from discussions in the prior chapters, the more turbulent the environment, the more frequently coordinates of the relevant vectors may have to be adjusted. By the same token, the more rapidly these adjustments must be made, the more rapidly analyses must be completed.

Three different frequencies are suggested in this chapter. The first is the annual analysis of R&D investment, particularly the origin amount, of top firms in the medical device industry. The strategic leadership of example Firm M have decided to move from the investment plane of midsize firms to that of the big league. Given the expected rate at which leadership of the top firms can change the total R&D budget, an annual analytic frequency appears reasonable. As Firm M gets closer to their target investment plane, the leadership may want to increase the frequency of these analyses.

The second frequency is the semiannual evaluation of R&D investments at pharmaceuticals Company D. This firm's leadership may be able to shift resources among therapeutic areas (not change the origin amount) fairly rapidly. Given the more labile situation facing Lucida, semiannual or even quarterly frequency appears reasonable.

The third frequency is the continual surveillance of high-risk discovery efforts in Biotech Company 1 and Company D. Signal strength of a rapidly advancing science frontier can change very quickly, thus justifying Lucida's watchfulness.

The actual frequency with which competitive position is analyzed must be a matter of careful judgment. The faster that signal strength can be estimated, the larger the window of opportunity for action (Figure 16). Better to err with too-frequent assessments than too-seldom.

Number of Competitors. The number of firms included in the competitive reference group is also a matter of judgment. For the sake of simplicity, above, I assumed that the relevant group for the medical device and pharmaceuticals companies included only a few firms with competing products *on the same R&D investment plane.* One could argue that much larger companies possess clear scale advantage and much smaller companies possess clear scale disadvantage. (Again, scale does not by itself guarantee success.)

In reality, if there are firms on different investment planes that have strategically important research efforts or development projects, the choice is clear: Analyses of competitive position should be conducted within therapeutic, product, or target areas, with separate analyses of research and/or development. What is illustrated by the example of Lucida NeuroSciences is that the relevant set of rivals emerged only after analyses of overall R&D *and* therapeutic area research investments. Another simplifying assumption in my discussion was that CNS *research* was more important to Lucida's competitive position than CNS *development.* In actual analyses, all such assumptions should be made explicit and tested.

TECHNOLOGY STRATEGY ARCHETYPES

The codifying methodology based on vector quantities defines three technology strategy archetypes by which internal consistency and competitive position can be assessed.[7] Remaining chapters build on these archetypes.

The archetypes are derived from the following simplifying choices. For any investment plane, divide the X and Y axes into equal segments. Designate the resulting segments on the X axis, "magnitude" or internal capacity, as *buy* (little or no in-house investment), *50-50* (roughly equal investments in-house and externally), and *make* (primarily in-house investment). Designate the resulting segments on the Y axis, "direction" or risk willingness of the relevant portfolio, as *low, moderate,* and *high*. Within each plane, three technology strategy archetypes can be drawn. I have termed these the *technology entrepreneur,* the *technology manager,* and the *technology pursuer* (Figure 26).[8]

Within each archetype, biomedical firms behave and invest similarly (i.e., they share a similar technology strategy). The *technology entrepreneur* represents companies narrowly focused on high-risk efforts with heavy investment in internal capacity. This is the most precarious position but also one in which potential rewards are greatest. Biomedical firms in this archetype are often new entrants, such as startup biotechnology companies; they can also be firms in which the strategic leadership have decided to change competitive position (discussed in Chapter Seven). The technology identity corresponding to the entrepreneurial archetype is *defining the leading edge.*

The *technology manager* represents firms with more balanced risk profile (more is invested in moderate- and low-risk efforts than the entrepreneur) and with roughly equal investment in both internal and external capacity. Biomedical firms in this archetype are likely to be the large, estab-

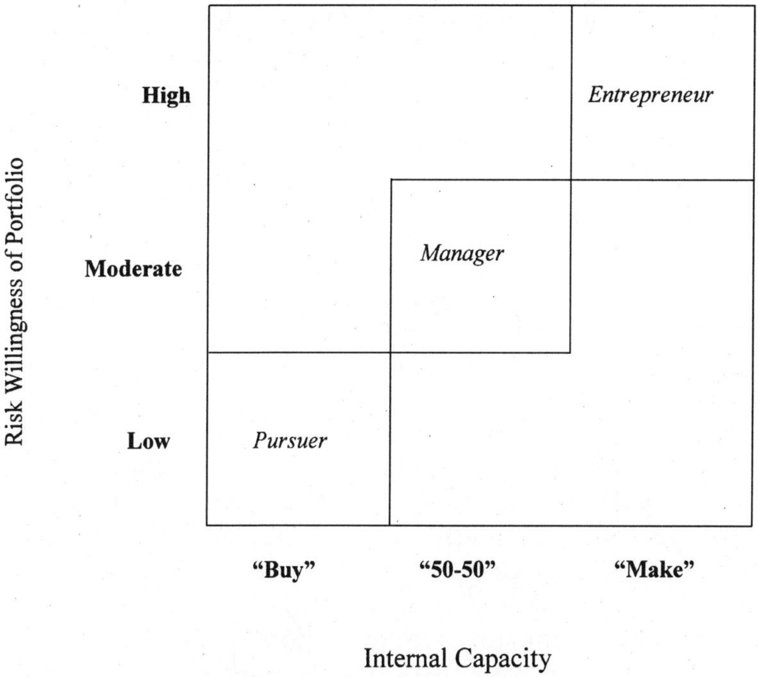

Figure 26. Technology strategy archetypes.

lished companies. The danger they face is in slipping "down," if investments in risk profile and internal capacity lag the competition. The technology identity corresponding to the managerial archetype is *remaining on the leading edge.*

The *technology pursuer* represents firms that emphasize low-risk efforts and rely heavily on licensing and joint ventures. Biomedical firms in this archetype may be companies that have slipped from one of the other two positions, because the leadership have waited too long to invest in core technologies in the emerging and early growth stages. They may also include firms new to the industry that must catch up. The technology identity corresponding to this archetype is *pursuing the leading edge*. The long-term viability of firms represented by the pursuer archetype is discussed in Chapters Six and Seven.

Like the framework and metrics from which they are derived, these archetypes are provided as additional strategic tools. I will use them in evaluating the internal consistency of the biomedical firm's science and technology investments (Chapter Six), and in preparing the firm to change competitive position (Chapter Seven). In the concluding discussion below, I give only a brief example of their utility.

The therapeutic area analysis of Lucida and its rivals, discussed above, reveals companies in each of the three archetypes (Figure 27). Biotech 1 corresponds to the *technology entrepreneur* within this investment plane. The firm has allocated 90% of CNS resources to internal capacity and 70% to high-risk (neurodegeneration, cognitive impairment, and/or autism) discovery efforts. Like all entrepreneurs, the firm is banking heavily on a successful outcome. If they fail, the company will likely not survive.

Company D and Lucida NeuroSciences correspond to the *technology manager* (or *quasi-manager,* as will be defined in Chapter Seven). Company D has invested slightly more than Lucida in internal capacity (75% versus 70%) and sizably more in high-risk discovery efforts (63% versus 40%). This puts Company D closer to the entrepreneur archetype at this level of analysis.

Biotech 3 corresponds to the *technology pursuer.* The firm has invested only 30% in internal capacity and 27% in high-risk discovery efforts. To put this another way, the firm has invested 70% of its CNS resources in external organizations and 73% in moderate- and/or low-risk efforts.

Biotech 2, on the other hand, falls outside the archetypes. The firm has invested 60% in internal capacity but

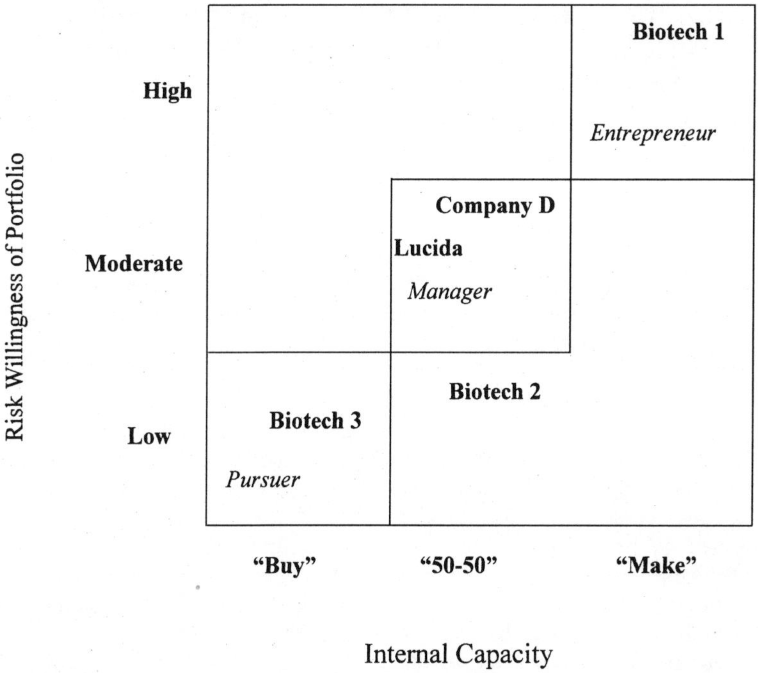

Figure 27. Use of archetypes to determine consistency and competitiveness.

only 23% in high-risk discovery efforts. As Figure 27 suggests, Biotech 2 could easily slip to the *technology pursuer* archetype, without an explicit decision by their leadership to do otherwise plus an appropriate shift of resources.

With regard to Lucida, the leadership are right to be concerned with the internal consistency of their investment. As I will discuss in more detail in Chapter Seven, at both the R&D and therapeutic area levels of analysis, Lucida's investment pattern conforms to the *technology manager* archetype, but *the firm is far too small to compete in the big league.* Consistency between Lucida's strategic vision and technology identity requires investment in a manner more consistent with the *technology entrepreneur.* How the firm might accomplish this is addressed in Chapter Seven.

NOTES

1. In this study (also cited in Chapter Four), the researchers observed how high-technology firms were investing in technology and asked managers on what basis they made their decisions (K. Clarke, D. Ford, and M. Swan, "Company Technology Strategy," *R&D Management 19* (1989); the citations are from pp. 227 and 215, respectively).
2. This is from J. B. Barney, "Beyond Individual Metaphors in Understanding How Firms Behave," in R. P. Rumelt, D. E. Schendel, and D. J. Teece (Eds), *Fundamental Issues in Strategy,* Harvard Business School Press, Boston MA, 1994, p. 67.
3. These comments are drawn from a working paper by D. J. Teece, G. Pisano, and A. Shuen, "Dynamic Capabilities and Strategic Management," Working Paper #90-8, Center for Research in Management, University of California, Berkeley (1990), p. 20.
4. An example of a publicly available source of industry strategic intelligence is Standard and Poor's Industry Reports, *Health Care Products and Services,* McGraw-Hill Companies (1995). It must be noted that one large firm, General Electric, has a market share of nearly 50% in magnetic resonance imaging (MRI) devices. The hypothetical midsize firms discussed in this chapter are working only in high-risk telemetry R&D and not on manufacturing MRIs.
5. For the pharmaceuticals sector, SCRIP's Annual Review Issues provide industry strategic intelligence as well (Surrey, U.K.).
6. The table is a brief excerpt of more than 100 diseases and conditions that were ranked independently by U.S. and Japanese pharmaceuticals scientists (A. Sapienza and I. Innami, "The Uncertainty of Pharmaceutical Research," presented at Technology Assessment and Forecasting Conference, Zurich, Switzerland (1993)).
7. Archetypes or postures are commonly described in, especially, the organization theory literature. Miles and Snow, for example, describe four organizational types that correspond to "a particular configuration of technology, structure, and process that is consistent with its strategy. These organization types . . . we have named the Defender,

the Reactor, the Analyzer, and the Prospector. . . ." Two of the technology archetypes I define are similar to the Miles and Snow types: *entrepreneur* is similar to the *defender,* and *pursuer* is similar to *reactor.* (R. E. Miles and C. C. Snow, *Organizational Strategy, Structure, and Process,* McGraw-Hill, New York, 1978.)

8. I first used vector metrics to evaluate the nine major Japanese pharmaceutical companies (A. Sapienza, "Assessing the R&D Capability of the Japanese Pharmaceutical Industry," *R&D Management* 23 (January 1993), 3–16).

CHAPTER SIX

CONSISTENCY AND COMPETITIVENESS

Because internal consistency between the strategic vision and how R&D investments are allocated is so crucial to the competitive success of the biomedical firm, the purpose of this chapter is to examine consistency and competitiveness in more detail. The chapter discusses the four major elements that must be consistent, examines consistency with Solar Imaging as case example, and describes some of the competitive perils of inconsistency, using illustrative firms from Chapter Five.

The methodology and archetypes presented in Chapter Five are important tools for assessing competitive position from the perspective of biomedical science and technology. But they are even more important for determining the internal consistency (or lack) of the firm's science and technology investments.

I believe the competitive necessity for the strategic leadership to ensure consistency among four major elements—strategic vision, technology identity, risk profile, and development of internal capacity—cannot be emphasized too strongly. Moreover, because of the turbulence of the environment, the leadership must continually assess whether these elements remain consistent, to ensure the biomedical firm's long-term competitive success.

Certainly, the problems of *in*consistency have been widely noted. One author defined technology capabilities broadly as a system with four components: knowledge and skills, physical technical assets, management, and values and norms. Not only gaps within each but also inconsistency among them rendered the organization vulnerable to competition.[1] Similarly, knowledge-creating companies (e.g., firms in the biomedical industry) were described by other authors as a "bio-function-type" of organization—in other words, an organic system. Inconsistency among components of this organic system was detrimental to the "health" of the firm and resulted in competitive impairment.[2]

Because internal consistency is so crucial to competitive success, my purpose in this chapter is to examine consistency in more detail. In the following section I discuss the four major elements that must be internally consistent. Then, I review these elements with Solar Imaging as case example. Finally, I describe some of the perils of inconsistency, using illustrative firms from Chapter Five.

CONSISTENCY: THE FOUR MAJOR ELEMENTS

Let me begin by explaining what consistency is *not*. The word does not imply either homogeneity or lack of difference. There must be a climate of challenge as well as sufficient intellectual diversity within R&D if the output is to be innovative.[3] In such a climate, there are bound to be—indeed there must be—differences of opinion.

"Consistency" connotes harmony and agreement, as well as steadfastness of purpose. Harmony and agreement *with what* is key, of course. In the new model of strategy, the ultimate "what" is the strategic vision, which must be widely shared and adhered to throughout the organization (a basic condition from Chapter Five). This first major element, discussed below, also includes level of strategic sophistication, domain, business objectives, technological network, and values (Figure 28).

First Major Element: Strategic Vision

As defined in Chapter Two, the strategic vision (endpoint of the trajectories) is an articulation of what the biomedical

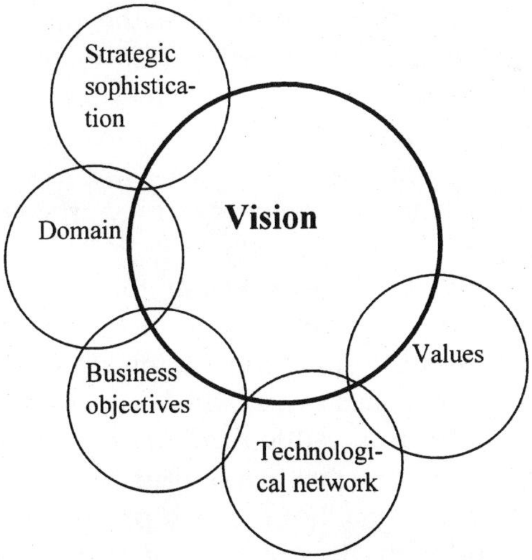

Figure 28. Internal consistency: the first major element.

firm could be in some desired future. That potential would encompass (among others):

- *competitive position*—to be among the top five firms in a particular sector; or to be number one in a niche, etc.
- *geographic scope*—to be global, international, etc.
- *treatment of constituents*—to be impeccably fair, to exceed external standards of quality, to contribute in a genuine way to communities in which the firm is located, etc.
- *technology identity*—the orientation towards science and technology that will characterize overall R&D.

Closely tied to the vision is the *level of sophistication* with which the leadership will approach technology strategy. In fact, these components have a chicken-and-egg relationship with each other, as implied in Chapter Two. The model of strategic trajectories presumes the highest level of strategic sophistication, which is to be instrumental and shape the future. But, in order to be instrumental, the leadership must have an ideal or vision towards which they strive (the "chicken"). Put another way, the leadership must be instrumental, in order to achieve the strategic vision (the "egg").

There is only one unique vision for each biomedical firm, but there are numerous ways for the organization to achieve its potential (termed the "doing" in Chapter Two). For example, corporate strategy, the blueprint for the business trajectory, determines the firm's *domain.* As defined in Chapter Four, domain specifies output (products and services), populations for whom the output is intended, and methods of producing the output. Corporate strategy also determines the firm's *business objectives,* such as target profitability, market share, sales growth rates, and so on. Both domain and objectives offer a myriad of choices. What is crucial is that every decision about domain and objectives be based on a positive answer to the question: *Will the choice help this biomedical firm achieve its strategic vision?*

Any disagreement about the "doing" must always be brought back to the strategic vision, the only meaningful yardstick by which the firm's progress and achievement can be measured. Using the above question as context, different choices that pass this first and most important screen can then be ranked according to such criteria as ease of implementation, cost, time to implement, and so on. Agreement about vision—in other words, consistency of decisions with the vision—is so important that some authors have contended that[4]

> . . . a major barrier to a firm's abilities to effectively anticipate changes in the marketplace is the lack of a shared vision about the firm's strengths and directions. . . . *[Shared] visions promote swift and smooth adjustments* (emphasis added).

The domain, as characterized above and discussed in Chapter Four, is embedded within a *technological network,* or set of general technologies that can be utilized by infrastructure as well as by research and development activities. In R&D as elsewhere, "technological network" decisions about technologies, scientific disciplines, and knowledge areas must also be consistent with the strategic vision ("will the choice help this biomedical firm achieve its strategic vision?").

Finally, all *doing* (all choices and actions resulting from those choices) reflects the *values* held within the organization. By "values" I do not mean published statements of what is espoused as important but rather what really are the determinants of behavior. Values either bring the organization closer to its potential (the vision) or drive it further away. There may be "vision-neutral" options offered by the domain, by business objectives, or by the technological network. *There are no vision-neutral values.* No one should be deluded that consistency between the strategic vision and the actual determinants of behavior is ever "good enough." Values must always be in complete harmony and agreement with the vision. If they are not, nothing else will be "good enough" for competitive success.

Second Major Element: Technology Identity

Discussion of the second major element (Figure 29) follows naturally from a discussion of organizational values, because values are the substrate of the biomedical firm's technology identity. *Values* that people in the organization hold influence their *attitudes* towards, among others, sci-

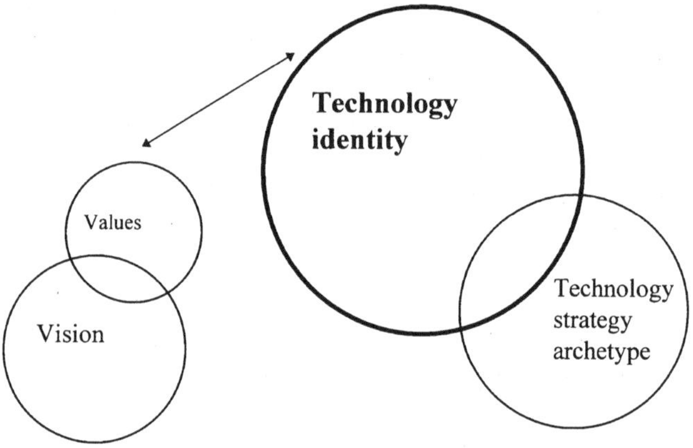

Figure 29. Internal consistency: the second major element.

ence and technology. As defined in Chapter Two, technology identity is the explicit choice of one of three general orientations or attitudes, which may be summarized as:

- defining the leading edge
- remaining on the leading edge
- pursuing the leading edge of relevant sciences and technologies.

One such orientation always distinguishes the biomedical firm's overall R&D efforts. The purpose of making the choice explicit is to evaluate and ensure its internal consistency with the strategic vision.

Interestingly, consistency between vision and values narrows the choice of technology identity. A vision (and the corresponding values) that involves being first in the CNS niche with state-of-the-art therapies requires an orientation (a technology identity) of defining the leading edge of the relevant sciences and technologies. A vision (and values) that involves being always among the top five firms in the medical device sector may be achieved with an orientation either of defining or of remaining on the leading edge of relevant sciences and technologies. Neither vision could be achieved with an orientation of pursuing the leading edge. But, inconsistency between vision and values almost invariably defines the latter identity. That is because the most likely inconsistency is between an ambitious vision statement and risk aversion as one of the actual determinants of behavior (i.e., values).

Before continuing with the other major elements, I want to make clear that technology identity and the corresponding technology strategy archetype should characterize *overall* R&D efforts over the *long term.* There are a number of circumstances in which the strategic leadership may decide to invest science and technology resources temporarily in a manner different from what is consistent with technology identity and archetype. The domain may have been changed and, therefore, the firm's "place" in the technological network may also have changed. As a result, the leadership may have identified other technologies in which they must invest more heavily. Or they may have selected future core technologies within the same domain and must shift investments in order to build competitive advantage. In these circumstances, it is appropriate to pursue the leading edge of "old" technologies.

The leadership may also decide to define the leading edge of certain technologies for a certain time in order to build the experience resource and create a competitive advantage. Consider that the leadership in medical device Firm M (Chapter Five) made the decision to invest heavily in high-risk efforts for that reason. (Their chosen technology identity is to remain on the leading edge.)

Over the long term, however, the manner in which the biomedical firm's overall science and technology resources are invested must be consistent with the technology identity and technology strategy archetype. In turn, the identity and archetype must be consistent with the strategic vision and (actually held) values. When decisions are made to invest differently (i.e., inconsistently), they must always be explicit and evaluated carefully on an ongoing basis. In other words, the leadership must ensure that "the choice still helps this biomedical firm achieve its strategic vision."

The third and fourth elements together represent the *experience resource* that must be built and continually developed to the biomedical firm's competitive advantage. Tangible R&D resources are important, but their competitive advantage lies not in their existence but in their use. As defined in the prior chapter, the potential experience advantage of the firm can be represented by the risk willingness of the relevant portfolio (the vector quantity of direction) and by the depth of in-house knowledge, skills, and capabilities (the vector quantity of magnitude). Each element is discussed below.

Third Major Element: Risk Profile

Over the long term, the risk profile (see Figure 30) of the overall R&D portfolio should be determined by, and remain consistent with, the chosen technology identity and

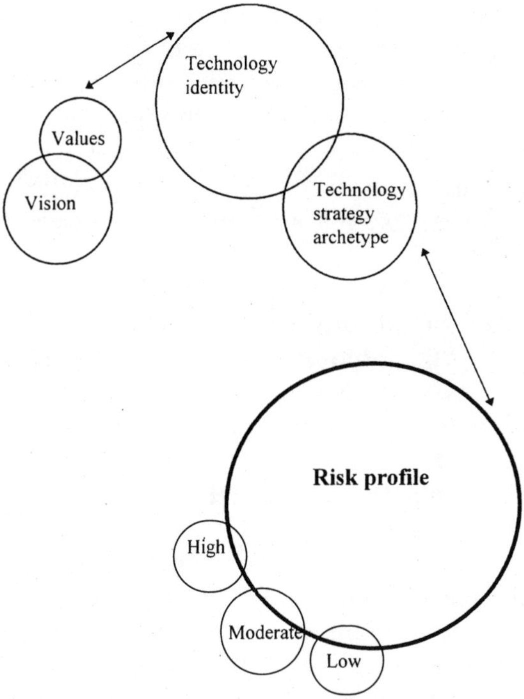

Figure 30. Internal consistency: the third major element.

technology strategy archetype. In other words, a larger proportion of science and technology resources must be invested in high-risk efforts to support an orientation of defining the leading edge of science and technology than are required to support an orientation of remaining on the leading edge. The smallest proportion is required to support an orientation of pursuing the leading edge.

As implied by the archetype definitions (Figure 26), the modal *entrepreneur* invests two-thirds or more of R&D resources to high-risk efforts, to define the leading edge. The modal *manager* invests R&D resources about equally in the three categories (or at least two-thirds in high- plus moderate-risk efforts), to remain on the leading edge. The modal *pursuer* (really a default position, as discussed later in this chapter) invests most of R&D resources in low-risk efforts. Like all general rules, these must be applied to the individual biomedical firm very thoughtfully. If the leadership decide to allocate resources somewhat differently from what is implied by the archetype, the reasoning should be made explicit and the consequences examined carefully.

For therapeutic (or target, or product) area efforts, at least two additional considerations influence what proportions of that origin amount should be invested in the three risk categories. One consideration is what fraction of total R&D investment is accounted for by the therapeutic area efforts. If the proportion is sizable, it makes sense for the risk profile to map that of the archetype.

The second consideration is the strategic importance of the therapeutic area at a given time. If competence in a particular area is needed quickly (e.g., to support movement from one domain or competitive position to another), it makes sense for the leadership to invest more heavily in high-risk efforts in that area than is the case for R&D taken as a whole. This was exemplified by medical device Firm M. On the other hand, if only minimal competence is needed for the short term, it makes sense to invest minimally.

Fourth Major Element: Internal Capacity

The last major element to be discussed (Figure 31) is the most critical. Numerous scholars have commented on the "the great weight played by the aptitudes of key individuals and by interpersonal networks" in competitive success.[5] After all, what is the biomedical organization if not the knowledge, skills, and capabilities of people within R&D (and other infrastructure functions)?

Simply put, building and developing the experience resource into a competitive advantage requires that invest-

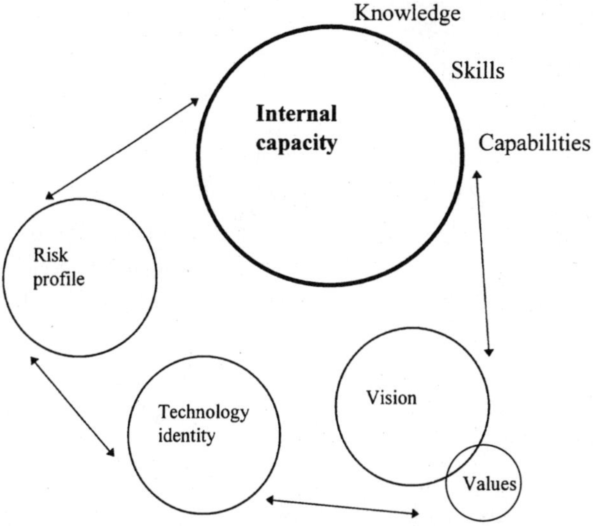

Figure 31. Internal consistency: the fourth major element.

ment in internal capacity match that of the selected technology identity and technology strategy archetype (which, in turn, dictate the risk profile). However, this investment must be understood to mean more than head count, although R&D head count was the basis for the vector analyses in the prior chapter. The most important investment is what goes *into* the head (count), to enhance organizational innovation. Highly trained specialists, with relevant practical experience, are vital to the biomedical firm's success.

Determining internal consistency between risk profile and the proportion of R&D allocated to internal knowledge, skills, and capabilities must be based on this expanded definition of investment. As an example, the leadership of an entrepreneurial firm would invest roughly two-thirds of the R&D budget in high-risk efforts to define the leading edge. That amount would be invested in both internal salaries and support for activities that would ensure that their scientists could define the leading edge, such as:

- collaborative efforts with public sector researchers connected to knowledge networks expected to define the leading edge
- travel to and/or short sabbaticals in facilities in which unusual technologies are employed, to expose scientists to a wider range of possibilities from the technological network
- cross-training and rotation to other areas of the firm, to discourage "silo" formation and narrowness of outlook.

These activities, and all organizational assistance needed (such as process consultation to foster challenge and foment within R&D), help to create and develop the experience resource into a true competitive advantage.

In Addition . . .

Before continuing with an illustration of these elements in a hypothetical biomedical firm, I want to comment on two additional issues of consistency. The first is consistency between use and advancement of technology; the second is consistency between (or alignment of) business and technology trajectories.

Technology Use, Technology Advancement. R&D resources, tangible and intangible, must be used to achieve business objectives ("technology use"). As I have discussed throughout this book, they must also be developed continually, to ensure future competitive advantage ("technology advancement"). As fast as the biomedical science frontiers advance, so too must the organization's technology assets and strengths. One observer stated that "even a constant rate of technological change implies that from one period to the next, the absolute amount of change—and the corresponding technological and managerial challenge—increases exponentially."[6]

The same principle that applies to current competitiveness, *internal consistency*, applies to future competitiveness. The leadership's efforts to advance technology must be consistent with vision and technology identity and at the pace appropriate to the relevant frontiers. There is, in effect, an investment plane corresponding to future capacity, and the same vector metrics should be used to determine internal consistency. Thus, if the (current and desired future) technology identity is to define the leading edge, two-thirds of "technology advancing" R&D resources should be allocated to high-risk efforts by in-house personnel. The magnitude of the "technology advancement" vector must be the same as the magnitude of the "technology use" vector; only the origin amount will differ.

Business and Technology Trajectories. My earlier discussion of strategic vision took for granted that (1) coordinates of the R&D vector were determined and set in accord with business objectives, and (2) coordinates of the business vector took into account the technology requirements. However, even before business objectives are agreed upon, there must be candid and intense discussions among corporate, R&D, and other functional managers about the implications of these objectives for short- and long-term science and technology resource allocation.

Inevitably, there will be times when a business objective, such as target profitability, cannot be achieved if sufficient resources are to be invested in R&D. Because meeting profitability targets brings rewards to senior leaders *now* but investing in R&D brings rewards years later, there is constant tension in the trajectory alignment process. Business and technology trajectories must be aligned (i.e., internally consistent) over the long term. But, aligning them will always involve short-term coordinate adjustments to the business vector and/or the R&D vector. This is illustrated in the following section and in Chapter Seven.

CONSISTENCY: AN ILLUSTRATION

Organizational life is messy. People do not sit down and, *de novo,* articulate their strategic vision (plus domain, business objectives, technologies, values) as the basis for subsequent R&D investment and proceed systematically and logically from there. The case of Solar Imaging is much more realistic and, therefore, a good illustration of the four major elements in a "real" organization.

Background

Solar Imaging, the hypothetical midsize medical device company, currently has revenues of about $1 billion and R&D investment of $80 million in two divisions. In 1990, as described in Chapter Two, corporate leadership had announced aggressive business objectives regarding profitability, market share, and revenue growth. To support these objectives, they changed the proportion of money invested in R&D in their two divisions.

Solar-Port ("Division X") was responsible for a range of portable X-ray devices and, in 1990, had a large market share in these products. However, industry analyses showed that sales growth rates had been declining steadily for the past 3 years, because newer technologies (e.g., ultrasound) were anticipated by the buyers. The leadership decided to reduce R&D investment in this division from its peak of 80% of total to about 30% in 3 years. This shift would, they believed, allow Solar-Port to continue to serve the current customer base, while providing increased investment in the new technologies expected to fuel company growth.

Solar-Link ("Division Y") had been created in 1987, to focus on the potential of telemetry linking any manufacturer's diagnostic imaging devices to centralized data bases. Although sales of this division accounted for a minute proportion of total company revenues in 1990, the leadership decided to increase R&D investment to about 70% of total in 3 years.

By the time of the analyses in Chapter Five, although the budget had been shifted from Solar-Port to Solar-Link in the above proportions, only 30% of company-wide R&D had been invested in industry-defined high-risk efforts. Why? R&D leadership in Solar-Port had entered numerous external alliances in the same, mature, X-ray technologies, to make up the difference in their budget. R&D leadership in Solar-Link, on the other hand, had underinvested in high-risk efforts (such as decision-support software), compared with the competition. They also licensed-in network technologies aggressively. As a consequence, only 19% of company-wide R&D was invested in internal capacity. The analyses in the prior chapter indicated that the length of Solar Imaging's R&D vector was 36 and was the smallest among the four competitors. Business objectives were far from being achieved, and the company was considered an attractive acquisition candidate, because of its available cash.

Solar Imaging's Strategic Vision

As in many organizations, it took a crisis (the acquisition scare) for the leadership of Solar Imaging to step back and consider what had gone wrong. Encouraged by the Board of Directors, and with some external assistance, they undertook a complete rethinking of the company's corporate and technology strategies.

Their first activity was to decide what they wanted Solar Imaging to look like in 20 years. They began, as they should, with the *being* (including treatment of constituents) and then discussed future competitive position. After much deliberation, people agreed that the company was too small to compete head-on with the big league firms like General Electric, and that a niche strategy was appropriate. But what niche?

Debates about their original growth targets led to agreement that this company should be risk willing (*values*) and that a focus on Solar-Link's output was desirable. The leadership concluded that their ideal was to be always among the top three firms in the global telemetry link niche (*vision*).

From this, the leadership could specify Solar Imaging's domain and place in the technological network in more detail: turnkey telemetry systems with data base and decision-support capabilities, future expansion to critical path generation software, and audit for selected disease management areas. The ambitious business objectives articulated several years earlier were retained.

The discussions of vision, domain, technological network, and business objectives were intense and arduous. But the most difficult were those involving organizational values. Again and again people had to come back to the now-articulated vision and talk about what behaviors were required to achieve it. As they reviewed the past several years' performance, they concluded that risk aversion was one of the actual determinants of their behavior.

Solar Imaging leadership recognized that everyone must behave in a much more risk-willing manner. Thus, as part of modeling the "new" values, whenever people proposed "covering their risks" by a sizable investment in lower-risk efforts ("just in case"), they were helped to confront the difference between *espousing* risk willingness and *being* risk willing.

This change was neither quick nor easy. The leadership in both divisions continued to argue about the need to balance investment and guard against failure—until they were brought back to the vision and its consequences. Consensus was long in coming (the better part of one year), but at last all agreed that they would support the vision, no matter how personally threatening it felt as they tried to achieve it. And they agreed that in the short-term they could *not* achieve a number of the aggressive business objectives if they were to create and develop the required knowledge, skills, and capabilities consistent with their strategic vision.

The Board of Directors was not pleased initially. After much discussion with R&D leadership, the Board realized that the delay in achieving certain business objectives was necessary if Solar Imaging were to survive in the turbulent health care environment. The R&D leadership were insistent in their belief that no biomedical firm could be competitive over the long term with what had been their *pursuer* technology strategy (this is addressed in the concluding section).

Solar Imaging's Technology Identity

As expected, discussions of values led to discussions of the orientation towards the relevant sciences and technologies that should distinguish overall R&D efforts. People soon concluded that they had really only one choice: to define the leading edge. Why? Their niche strategy, desired competitive position, selection of domain, and business objectives could only be attained by that choice over the long term. In the near term, the allocation to Solar-Port would be inconsistent with that identity, because the division was committed to alliance agreements for several years. But, as much as possible, R&D leadership in Solar-Port would attempt to seek early termination of alliances that kept them from high-risk and in-house efforts. They would also assess spinning off or selling off the low-risk projects as appropriate.

Solar Imaging's Experience Resource

Solar Imaging had invested 30% of company-wide R&D in high-risk efforts and 19% of that in-house. In the context of their vision and technology identity, the leadership determined that the ideal allocation was two-thirds of overall R&D to high-risk efforts by in-house scientists, in accord with the entrepreneurial archetype. Solar-Port would systematically shift from X-ray to ultrasound devices (moderate risk); Solar-Link would continue to focus on high-risk telemetry.

Together, R&D leaders from Solar-Port and Solar-Link discussed what the company-wide vector metrics should be and what activities were required to achieve them ("technology use"). They also planned carefully for future competitive viability, by ensuring that "technology advancing" resources would be invested in the proportions consistent with an identity of defining the leading edge.

R&D scientists from both divisions were very excited about the company's decisions (remember that, earlier, several key scientists had left in frustration), despite the need for most of Solar-Port scientists to retrain extensively. During their many discussions, they identified a number of public-sector laboratories working on topics that could be crucial to Solar Imaging's future experience advantage. The leadership agreed to support collaborations and to rotate a number of Solar's discovery scientists through these laboratories, to gain important new skills. An in-house seminar series was also planned by the scientists, to expose as many as possible to technologies that might be used in new product development (i.e., new technologies from the technological network). In addition, R&D leaders of both divisions decided to ensure challenge and foment by seeking candid feedback on their own behaviors. To many, this was the most personally threatening (and risky) activity they had undertaken thus far, but they agreed it would help them achieve their vision.

Postscript

Figure 32 summarizes this case example. How might the story end? Solar Imaging may not have time to reorient itself and achieve internal consistency. Remember that Firm M is in the strongest competitive position with regard to these midsize medical device firms (Chapter Five). On the other hand, the leadership in Firm M is attempting to enter

Vision	Always among top firms in niche Telemetry systems (data base, decision support; critical path generation and audit; disease management) Risk-willing
Technology Identity	Define leading edge of sciences and technologies relevant to niche Entrepreneurial technology strategy archetype
Risk Profile	Over long term, two-thirds total R&D to high-risk efforts
Internal Capacity	Over long term, two-thirds total R&D allocated internally, including • External collaboration • Rotation of scientists • Exposure to technologies • Challenge and foment

Figure 32. Solar Imaging's internal consistency.

the big league. If Solar Imaging leaders continually assess the internal consistency of their investments and make the appropriate R&D vector-coordinate adjustments, the firm just might succeed in its chosen niche.

Leaving Solar Imaging in far better (hypothetical) shape, I describe some of the perils of inconsistency, using other illustrative firms from Chapter Five.

THE PERILS OF INCONSISTENCY

As alluded to in the above discussion, inconsistency is perilous *because of* the turbulent space through which the trajectories must pass. If the health care environment were stable, internal inconsistency would leave the biomedical firm vulnerable to competition, but inconsistency would not in itself be perilous (i.e., with imminent threat to survival). Inconsistency always implies suboptimal efficiency and, therefore, excess cost.

Under turbulent conditions, however, the strategic leadership has a much shorter time to correct inconsistencies that leave the firm vulnerable. And, while there are inconsistencies under turbulent conditions, the technology trajectory is more likely to be knocked off course (leaving the organization even more vulnerable). Finally, correcting trajectories that are far off course requires a longer time and higher cost. It is because of turbulence that internal consistency is so important.

Three dynamics that are partial causes of high turbulence in the health care environment have an enormous impact on the biomedical industry:

- cost-constraining pressures from strong buyers (the microeconomic dimension), coupled with stringent requirements that products demonstrate clear advantages over alternatives (i.e., their benefit exceeds their cost)

- rapidly advancing science frontiers, compressing product and process life cycles (the scientific and technologic dimension)
- consolidation of firms, increasing the scale of resources needed to remain competitive.

These dynamics *reduce the degrees of strategic freedom* available to the biomedical firm (cf. Figure 2), primarily the choice of technology identity and technology strategy archetype. The necessity to demonstrate clear cost advantages coupled with rapidly advancing science frontiers make it impossible for any biomedical firm to be competitive over the long term with a technology identity of *pursuing* the leading edge. This was the argument made by Solar Imaging leadership to their Board of Directors, in support of the chosen technology identity of defining the leading edge and the entrepreneurial archetype.

Consolidation of firms, with the attendant increase in the scale of resources needed to remain competitive, also reduces the degrees of strategic freedom. Only the largest pharmaceuticals and medical device companies will be able to compete successfully as *technology managers*. That is, the scale of these firms will permit a broad domain (i.e., applying a critical mass of scientists, equipment, and funding to a number of therapeutic or product areas), with sufficient investment across the three risk categories, and use of a balance of internal and external capacity. (The modal firm in this archetype allocates at least two-thirds of R&D resources to moderate- plus high-risk efforts, with about half of that to in-house capacity.) Only in these largest firms is the strategic leadership able to make a choice between a technology identity of defining or of remaining on the leading edge.

For small and midsize firms like Solar Imaging and Lucida NeuroSciences, it may be impossible to survive with other than a niche strategy (focused domain). Some have even argued that smaller companies can only be successful as niche players, making use of emerging technologies and with "managerial willingness to risk their available resources on the exploration of new therapeutic niches":

> It is important to emphasize the market volatility of the therapeutic niche: when the required core technological competence is at the embryonic or emerging stage, only a few contenders have the capability to explore the niche by addressing unmet medical need. Later, as the technology matures, it will be widely used and the attractiveness of the niche will decline.[7]

Of course, scale by itself does not insure the largest firms against failure. As I said earlier, the competitive advantage of R&D resources lies not in their existence but in their use. The downside of large scale includes bureaucracy, creeping risk aversion, and loss of ability to remain on, let alone define, the leading edge. For the biggest firms, success can breed arrogance. Referring to the largest firms in the pharmaceuticals sector, other authors contend that bureaucracy exacerbates the difficulty of keeping up with rapidly advancing frontiers:

> Corporations invest large amounts of expensive staff time, and many dollars, to develop often rigid, long-term R&D strategic plans based on conservative market research, reflecting yesterday's or today's knowledge.[8]

The largest firms must always be wary of competition from successful niche companies. Niche companies, the *technology entrepreneur,* can take sizable market share away from the *technology manager,* with products that provide therapeutic and cost advantages. This may be another factor influencing large firms to partner with niche companies (such as pharmaceuticals and biotechnology companies), to "coopt" the competition.

In the concluding section, I discuss two perilous inconsistencies: (1) the inconsistency between vision and technology identity, and (2) the inconsistency between risk profile and internal capacity (i.e., inconsistency within the experience resource). Again, these inconsistencies are perilous because of turbulence and the three industry dynamics described above.

Vision Versus Technology Identity

One of the hypothetical medical device firms discussed in Chapter Five, Firm L, illustrates the perilous inconsistency between vision and technology identity. The strategic leadership apparently believe the firm can succeed within the *technology manager* archetype. However, the scale of R&D is too small in comparison with top firms (origin), their investment in internal capacity (40%) is less than the modal firm in that archetype (magnitude), and less than one-quarter of their overall R&D resources is invested in high-risk efforts (direction)

Firm L is, in fact, close to the *pursuer* archetype, with the following consequences:

- Because of the underfunding of high- and moderate-risk efforts, the resulting products are unlikely to offer sufficient cost-benefit to make them attractive to strong buyers.
- With declining revenues from less-attractive prod-

ucts, the leadership will be hard pressed to fund R&D at the same proportion of sales.
- With less available to invest in high-risk and in-house efforts, the leadership may become more risk averse.

This downward spiral may "seal their fate" as a pursuer. The stability of the pursuer position is extraordinarily dangerous,

> . . . not only because it is virtually impossible to move away from it (at least not without the infusion of considerable energy) but also because of the perfect fit between the core technology (mature) and risk attitude (averse). [Such] companies are easy targets for hostile takeovers. . . .[9]

Lucida NeuroSciences illustrates the same inconsistency. Their leadership had earlier decided that a niche strategy was appropriate to the firm's scale, but the allocation of R&D resources was more similar to the *technology manager* (48% to high-risk efforts, 52% of that in-house). Although Lucida is not in as dangerous a position as Firm L, this inconsistency is still perilous. How the leadership might change Lucida's competitive position is discussed in the next chapter.

Risk Profile Versus Internal Capacity

Even with consistency between vision and technology identity, there can remain a second perilous inconsistency between risk profile and internal capacity—that is, inconsistency within the experience resource. The strategic leadership must ensure that R&D resources are appropriately allocated to the three categories of risk *and* to internal knowledge, skills, and capabilities. There have been a number of examples of this second perilous inconsistency among the firms discussed in Chapter Five.

For instance, of the medical device companies, Firm P exhibits a marked inconsistency between risk profile and internal capacity. The leadership had decided to change the technology identity from remaining on to defining the leading edge, which is appropriate to the firm's size and a niche strategy. But they have invested 57% of R&D in high-risk efforts—less than needed to define the leading edge, although they are increasing this—with only 21% of that invested in-house.

What has happened in Firm P is not uncommon and may reflect a decoupling of the licensing strategy from the technology strategy. Data base and decision support technologies were licensed-in very aggressively, to support the change in technology identity. At the same time, however, the leadership failed to realize that, by doing so, they were essentially outsourcing their core technologies (Chapter Four). The firm is in a perilous position because of this. Knowledge, skills, and capabilities in-house will be very weak in comparison with other *technology entrepreneurs.* The leadership may not have time to build up the required expertise, and while they are trying to do so, the firm may be knocked off course.

Lucida NeuroSciences exhibits this second inconsistency but from a different perspective. Within this firm's major therapeutic area, central nervous system therapies, the leadership has invested too little in high-risk efforts (40%). The investment in in-house personnel, however, is appropriate to the *entrepreneur* archetype (70%). As discussed in Chapter Five, Lucida is facing strong competitive pressures from two companies whose scientists may succeed in putting a therapy into clinical trials earlier than Lucida scientists.

In terms of both overall R&D and CNS therapeutic area investments, the leadership in Lucida must focus more narrowly on high-risk efforts. Like Firm P in the medical device industry, they may not have time to make these changes, and while they are trying to do so, the firm may be knocked off course.

IN SUMMARY

I believe that, because of the high level of turbulence in the health care environment and because of cost constraints, rapidly advancing science frontiers, and consolidation of firms:

- No biomedical firm will be able to compete over the long term with a technology identity of pursuing the leading edge.
- In only the largest biomedical firms will the leadership be able to choose between a technology identity of remaining on or of defining the leading edge.
- The leadership in small and midsize biomedical firms may have no option but to select a niche strategy (focused domain), the *technology entrepreneur* archetype, and the technology identity of defining the leading edge.

With regard to other aspects of domain, therapeutic area, technologies, business objectives, and so forth, there are many degrees of strategic freedom. But, turbulence and the three dynamics, above, greatly reduce the identity and archetype options available to the biomedical firm.

In the following chapter I again address the subject of competitiveness, but from the perspective of changing competitive position. *I assume that internal consistency among the four major elements can be achieved.*

NOTES

1. The components of the "technology system"—knowledge and skills, physical technical assets, management, and values and norms—are very similar to the four elements in my model that must be consistent. The work is by D. Leonard-Barton, *Wellsprings of Knowledge,* Harvard Business School Press, Boston, MA, 1995.
2. Another work that addresses consistency is I. Nonaka and H. Takeuchi, *The Knowledge-Creating Company,* Oxford University Press, Oxford, England, 1995. The quotation is from p. 82.
3. I have emphasized the need for intellectual challenge and foment within R&D in *Managing Scientists,* Wiley, New York, 1995.
4. Strong corroboration of the importance of strategic vision and the need to link vision with technology strategy is found in R. A. Goodman and M. W. Lawless, *Technology and Strategy,* Oxford University Press, New York, 1994. The quotation is from pp. 89–90.
5. This quotation is from P. Adler, "Technology Strategy: A Guide to the Literatures," Unpublished report, Stanford University Department of Industrial Engineering Management, Stanford, California (1988), p. 40.
6. Ibid., p. 6.
7. A very provocative article on the niche player (entrepreneurial firm) is that of S. Jorgensen, I. Jensen, and M. Edwards, "Surviving as a Niche Player in the Pharmaceutical Industry," *Drug Development Research 30* (November 1993), 112–120. The quotation is from p. 115.
8. A related discussion is provided by J. A. Weisbach and W. H. Moos, "Diagnosing the Decline of Major Pharmaceutical Research Laboratories," *Drug Development Research 34* (March 1995), 243–259. The quotation is from p. 250.
9. The quotation is taken from the Jorgensen, *et al.*, article (Note 7, above), p. 114.

CHAPTER SEVEN

CHANGING COMPETITIVE POSITION

A key requisite for changing the competitive position of the biomedical firm is the ability to achieve consistency among vision, technology identity, risk profile, and internal capacity (see Chapter Six). Assuming that such change is required for long-term survival, the purpose of this chapter is to provide a framework and illustrations of changing competitive position by *moving to a new technology strategy archetype.* The chapter briefly reviews what is and what is not involved in moving to a new archetype, describes a framework of four basic steps, and uses hypothetical firms to illustrate appropriate changes from one competitive position to another.

MOVING TO A NEW ARCHETYPE (WHAT IS, WHAT IS NOT, INVOLVED)

Being able to achieve consistency among vision, technology identity, risk profile, and internal capacity is a key requisite for successfully changing the competitive position of the biomedical firm. From the perspective of science and technology, changing competitive position means moving to a new technology strategy archetype. By definition, moving to a new archetype requires changing the firm's technology strategy.

Moving to a new archetype may involve modification of domain and business objectives and of the strategic vision as well. For two of the three changes in competitive position described in this chapter, archetype change also involves a change in technology identity.

From the perspective of science and technology, what is not involved in changing competitive position is all the activity required to become more competitive. I assume that everyone in the biomedical firm is working to reduce cost, to improve quality, and to shorten throughput times. And I assume that the leadership are making coordinate adjustments as needed, within the same archetype. That activity, *becoming more competitive,* is enormously important but is taken for granted in this book.

Changing competitive position follows a detailed analysis of the health care environment (Chapter Three) and assessment of the biomedical firm's current competitive position (Chapter Five). Because of turbulence and the dynamics discussed in Chapter Six—cost constraints, rapidly advancing science frontiers, industry consolidation—the degrees of strategic freedom are limited. Thus, in this chapter I focus on what I believe are the three appropriate competitive position, technology archetype changes (Figure 33):

- from *quasi-manager* (defined below) to *entrepreneur*
- from *entrepreneur* to *manager*
- from *pursuer* to *manager.*

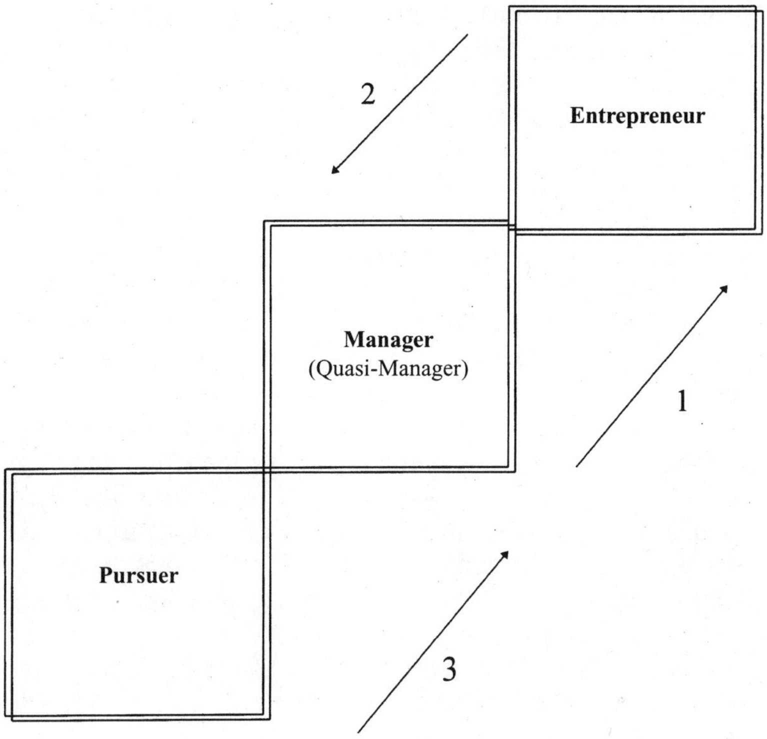

Figure 33. Appropriate archetype changes.

The first move is appropriate for small and midsize firms, like Lucida NeuroSciences, that have invested R&D resources as if they were a *technology manager* but cannot, because of their scale, compete against the largest companies in the relevant sector. (That is why they are labeled "quasi-manager.") This move requires greater risk willingness and the desire to be all that the entrepreneurial archetype implies.

The second move, from entrepreneur to manager, is appropriate for diversified companies with a successful biomedical division, whose leadership want to expand the biomedical business. Because of the firm's diversification, there are assets that can be sold to increase the scale of biomedical R&D, so that it will encompass the breadth and depth of the *technology manager.* Because of the division's success, there is a good basis for such an increase in scale. The move requires explicit acknowledgment of, and preparation for avoiding, the problems of expansion ("rigid, long-term R&D strategic plans based on conservative market research, reflecting yesterday's or today's knowledge").[1]

The third move is appropriate for large firms that have lost their technology advantage and "slipped" to a lagging position. These organizations have the scale to compete as *technology manager.* What is required is a change in values, as well as commitment to shift R&D resources to more risky efforts and to build up internal capacity appropriately.

Other changes in archetype are possible but either not reasonable or very unlikely:

- from *entrepreneur* or *manager* to *pursuer*
- from *pursuer* to *entrepreneur*
- from (genuine) *manager* to *entrepreneur.*

The first move, from entrepreneur or manager to pursuer, is not reasonable. Given the turbulence and dynamics noted earlier, this move is tantamount to exit from the biomedical industry. The second move, from pursuer to entrepreneur, is also not reasonable. Such a change could not be accomplished within a short enough period of time for the biomedical firm to survive the transition. In addition, this move would probably require so many new personnel that, for all intents and purposes, the former pursuer would no longer exist. A large, lagging biomedical firm could spin off a smaller company intended to be a niche player. In this case, the discussion of the move from quasi-manager to entrepreneur is applicable. The third move, from manager to entrepreneur, is very unlikely. Leadership of a successful, large firm would be hard pressed to justify breaking up the organization into smaller niche companies. (It is possible that one or more divisions might be spun off as entrepreneurs.)

In the following section I describe a simple framework, four general steps that are always involved, for implementing an appropriate archetype move. In the concluding section, I use three hypothetical firms (including two discussed earlier) to illustrate each change. Let me be clear that, because of the enormity of the undertaking, I do not recommend changing competitive position unless the biomedical firm's long-term survival is at stake.

A FRAMEWORK FOR CHANGE

My framework for changing competitive position takes as given that the leadership (1) have carefully evaluated the biomedical firm's current competitive position and (2) are able to achieve consistency among the four major elements discussed in Chapter Six. I assume that the impetus for change is realization that the firm's survival depends on moving to a new archetype. The four general steps in the framework for change are: (1) selection of appropriate, future, core technologies; (2) reallocation of R&D investments; (3) development of the experience resource; and (4) business vector coordinate adjustments. Each is described below.

Selection of Future Core Technologies

The process of evaluating current competitive position generally reveals different aspects of the domain (products and/or services; populations; discovery, development, production technologies) that the leadership believe will be important to future success. As I discussed in Chapter Four, modification of domain will also change the biomedical firm's "place" within the technological network. Thus, moving to a new archetype begins with the identification of the widest range of potential technologies from that network. The leadership's objective is to select future core technologies that will best support the new domain and archetype.

Of all possibly relevant technologies in the network, only those for which the following three questions can be answered affirmatively should be selected:

1. *Are these technologies a* best match *with existing knowledge, skills, and capabilities, to leverage the firm to the new archetype quickly and effectively?*

Every organization, even one that falls within the pursuer archetype, has unique competencies in discovery and development technologies. Because of the turbulence of the environment, these must be used to leverage the biomedical firm quickly and effectively from the old to the new archetype. Of the many technology options available, the leadership must initially select those that best match existing (i.e., demonstrated) knowledge, skills, and capabilities.

Hiring or acquiring completely new competencies is a dangerous tactic under conditions of turbulence and competitive vulnerability of the firm. Certainly, as I discuss in the following chapter, new competencies must be continually developed at a pace consistent with the advancing frontiers. But, hiring or acquiring new competencies must also be at a pace at which the people can be fully integrated into R&D culture as well as operations. It is the potential for "centrifugal" actions that weighs heavily against the tactic of buying competency the firm does not have in-house.

2. *Given best match with the firm's current internal capacity, are these technologies within the* area of technology advantage?

Analogous to the area of competitive advantage, in which signal strength is weak and cost of response is low (Chapter Three), there is an area of technology advantage, as illustrated in Figure 34. The leadership must identify, from the subset of "matching" technologies in the network, those that are also in the *emerging* and *early* (not late) growth stages of their life cycle. Weak signals of uncertainty and equivocality in these stages will include (cf. Chapter Four):

- sparse published information; information is generally word-of-mouth within several scientific circles
- few published patents (if applicable)
- experience restricted primarily to academic and public-sector institutions
- scarce cost data
- multiple and conflicting opinions about the use, role, and scope of the technology.

When signals are weak, the cost of developing capacity in the technologies is lower (hence, this is an area of technology advantage). As signal strength increases, so too

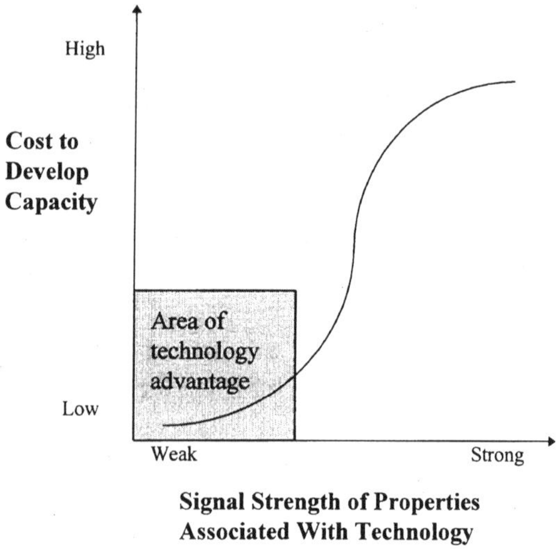

Figure 34. Area of technology advantage.

does cost (Figure 34). Large, established firms can afford to invest in future core technologies in the late growth stage of the life cycle. For leaders desiring to move a small or midsize biomedical firm to the entrepreneur archetype, recognizing the area of technology advantage (emerging and early growth stages) is particularly important. This is the only area in which their firm can afford to build expertise.

3. *Given best match and area of technology advantage, will these technologies help the firm to achieve its* strategic vision?

As I emphasized in Chapter Six, every choice must be made in the context of the strategic vision. Of the technologies that meet the first two criteria, only a subset will also be a best fit with the vision. Selection of future core technologies must always be made in the light of that shared ideal.

This first step is summarized in Figure 35.

Reallocation of R&D Investments

Once future core technologies are selected, investments must be shifted according to three general reallocation rules. First, for "old domain" technologies in the mature phase (i.e., noncore), R&D investment must be reduced im-

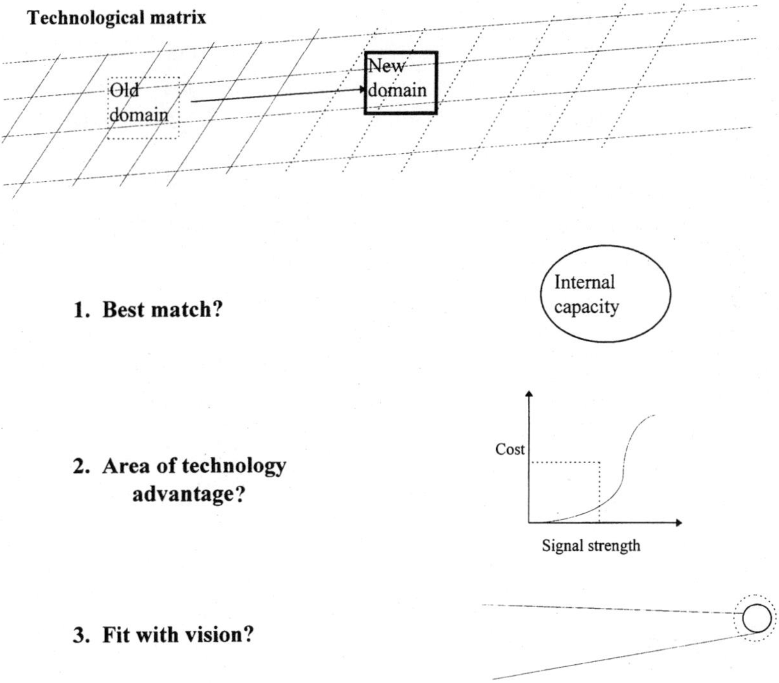

Figure 35. Future core technology section.

mediately to free up resources. The leadership should determine the minimum allocation needed to support the existing customer base and devise a reduction plan that least disrupts the relevant activities.

Second, for "old domain" core technologies that are not in the mature phase but will be superseded by new technologies, the decision is different. The leadership should shift investment so that levels of uncertainty and equivocality (and risk) within R&D are consonant with the levels in competitors' organizations. In other words, the investment required is that which will maintain competitive position (cf. Figure 18) during the transition to the new archetype. The leadership must carefully evaluate strategic intelligence on an ongoing basis, so that they neither over- nor underinvest in what are becoming "old" technologies.

Third, for selected future core technologies, the leadership should shift investment so that the levels of uncertainty and equivocality (and risk) are reduced faster within R&D than they are within competitors' organizations. Their objective is to ensure that scientists develop new knowledge, skills, and capabilities as fast as possible, to provide competitive advantage (discussed below).

The size of the monetary investment in selected future core technologies must be consistent with the investment plane of the desired competitive position. The proportions of that new origin amount allocated to high-, moderate-, and low-risk efforts must map the proportions dictated by the new archetype. This will be illustrated in the concluding section of the chapter.

Development of the Experience Resource

The tangible resources represented by monetary investment in the new technologies must be built into a unique experience resource, in accord with the desired archetype and technology identity.

Simply ensuring consistency within an archetype is difficult enough. But, developing the experience resource while simultaneously moving the firm to a new archetype is even more arduous. During the transition, the biomedical firm must continue to operate effectively in its former competitive position. Late-stage development projects that will provide crucial resources should be brought to market as fast as possible. Because of the inevitable disruption that change brings, a number of R&D staff should be "walled off" to bring these projects to completion. When the projects are finished, staff must be quickly and smoothly reintegrated into the changing organization.

Programs and projects "upstream" in R&D must be rapidly moved forward, stopped, or outsourced, depending on their fit with the new archetype. If they are to be moved forward, again, staff must be shielded from the disruption of change and then reintegrated. If they are to be stopped or outsourced, morale issues must be confronted and scientists transferred as fast and as smoothly as possible to work on the future core technologies. This is a situation in which agreement about vision is most important; otherwise, there will be no "swift and smooth adjustments" to the new archetype.[2]

At the same time as these activities are taking place, new scientists will undoubtedly be recruited to supplement internal capacity. They, too, must be integrated into ongoing operations. Finally, existing staff must be supported in their efforts to:

- connect with new knowledge networks relevant to the new technologies
- become proficient in new skills
- acquire new capabilities.

As I stated above, the proportion of the new origin amount invested in building internal capacity should match the proportion dictated by the desired archetype. Again, this will be illustrated in the concluding section of the chapter.

Business Vector Coordinate Adjustments

All functions in the biomedical firm will be affected by the new archetype, not only R&D. While coordinates of the R&D vector are being changed by the first three steps, business vector coordinate adjustments must be made as well. For example:

- Marketing managers must consider the resources needed to market new products and/or services and the capability required to support them over their expected market life.
- Manufacturing and engineering managers must consider the equipment and production capacity required for new products, so that plants come on-stream at the appropriate times.
- Managers in all infrastructure functions—quality, regulatory, finance, etc.—must determine the resources needed to support R&D in the new archetype.

The Model and Reality

Changing competitive position is an enormous undertaking. What I have summarized in the above four general steps is, in fact, a multiyear, resource-intensive, highly disruptive activity. Although I will illustrate the three applicable archetype changes below, let me emphasize that I do not recommend changing competitive position unless it is absolutely necessary for the biomedical firm's survival.

Two of the three archetype moves—from *quasi-manager* to *entrepreneur,* and from *pursuer* to *manager*—are predicated on the fact that long-term competitive success is impossible otherwise. The quasi-manager cannot compete with the scale of the genuine manager; the pursuer cannot compete at all. Expansion of an entrepreneurial division of a company as the basis for moving to the *technology manager* archetype is the only change in competitive position that is strategic as opposed to a survival choice. Although less difficult to implement on some counts than the other two changes, it is still an enormous undertaking.

I believe it is far better for the biomedical firm if the leadership ensure that changing competitive position will not be required. If the necessity for change arises either from internal inconsistency (e.g., the quasi-manager) or from lack of awareness of turbulence and the dynamics of the health care environment (e.g., the pursuer), then the leadership have been remiss. The firm's science and technology resources have been allocated inappropriately, jeopardizing the organization. As implied in Chapter Six, the health care environment is harsh and unforgiving of such mistakes.

Figure 36 summarizes the steps involved in changing competitive position. Below, I use three hypothetical biomedical organizations (including two discussed earlier), to illustrate these steps for each appropriate archetype move: (1) from *quasi-manager* to *entrepreneur,* (2) from *entrepreneur* to *manager,* and (3) from *pursuer* to *manager.*

1. **Select future core technologies**

2. **Reallocate R&D investments**
    - Reduce investment in "old domain," noncore capacity immediately
    - Maintain position in "old domain" core capacity during the transition
    - Gain competitive advantage in future core capacity as fast as possible

3. **Develop the experience resource**
    - "Wall off" staff to complete late-stage projects; move forward, stop, or outsource upstream programs and projects as appropriate to new archetype
    - Recruit and integrate new staff
    - Develop new knowledge, new skills, new capabilities throughout R&D

4. **Adjust business vector coordinates as needed**

Figure 36. Framework for change.

CHANGING COMPETITIVE POSITION: THREE ILLUSTRATIONS

From *Quasi-Manager* to *Entrepreneur*

As should be clear from the prior two chapters, Lucida NeuroSciences is an example of a firm in which science and technology resources have been invested as if the organization were a *technology manager.* In reality, the company lacks the scale to compete with the largest organizations in its sector (pharmaceuticals). As portrayed in Chapter Six, Lucida also exhibits the two perilous inconsistencies: (1) inconsistency between vision and technology identity, and (2) inconsistency between risk profile and internal capacity.

Realizing that competitive success depends on their changing from this position (cf. Chapter Two), the leadership had proposed a niche strategy with an emphasis on central nervous system (CNS) therapies, thus reducing the number of therapeutic areas in which their scientists would work. Included in the vision for Lucida is always being among the top three firms in the chosen niche; the technology identity is to define the leading edge, and there are ambitious business objectives to be met. Despite agreement with the niche strategy, however, Lucida's science and technology resources have been invested as shown in Table 13.

Based on analyses of the firm's current competitive position, the leadership are uneasy about the proportions listed in the table. Nearly equal amounts have been allocated to research in each risk category (high = CNS, moderate = CV, low = AI). Within their major research area of CNS, only 40% has been allocated to high-risk neurodegeneration, cognitive impairment, and/or autism discovery programs (Table 11). Slightly more than half (52%) of overall R&D has been allocated to internal capacity; the proportion is higher within CNS research (70%).

Table 13. Lucida R&D Budget

Research Programs ($75 million)	Line Item Budget
Central nervous system (CNS)	$25 million
Cardiovascular (CV)	$30 million
Antiinfective (AI)	$20 million
Development Projects ($175 million)	**Line Item Budget**
CNS (Phase III)	$95 million
CV (Phase II)	$10 million
AI (Phase III)	$70 million

Implementing a change in competitive position involves, first, selection of future core technologies. In reviewing the options with their scientists, R&D leaders decide that gene signaling technologies meet the three criteria defined earlier. Because of the inherent volatility of all therapeutic niches, they decide to explore other emerging-stage technologies. Based on the work of an in-house task force, with outside expert opinion, the director of research recommends investing in molecular biology and gene technologies relevant to atherosclerosis and ischemia. The technologies selected are judged to meet the three criteria of match with internal capacity, area of (cost) advantage appropriate to the firm's resources, and fit with the strategic vision for Lucida.

Having selected future core technologies, and the leadership understand that technology assessment is an ongoing process, R&D investments must be reallocated. Painstaking review of the portfolio suggests that the most obvious set of programs to be stopped is antiinfective (AI) research. To provide support to the Phase III AI development projects (Table 13), the leadership agree to preserve $3 million of the $20 million in that therapeutic area research. The $70 million invested in clinical trials is a sunk cost; however, a marketing partner will be sought, to reduce Phase IV expenses.

The Phase III CNS development projects target neurodegeneration and use technologies that are important to Lucida's current competitive position. The leadership believe that the company could begin receiving revenues from these therapies, assuming successful completion of the clinical trials, within 3 years. They decide to invest more resources to the latter efforts, from the $17 million freed up by stopping AI research programs. Scientists recommend that a sizable amount of money be allocated to advanced information technologies that will enable Lucida to design and monitor all future trials more efficiently.

Of the $30 million invested in cardiovascular research, about $20 million are deemed to be allocated to efforts relevant to atherosclerosis and ischemia; the remainder is invested in "old domain" programs. The Phase II CV development project is a therapy for atherosclerosis and will be moved ahead as fast as possible, with a modest increase in investment.

In summary, the leadership determine that, on paper, they have $27 million to reallocate more appropriately as a *technology entrepreneur* ($17 million from AI; $10 million from CV). They consider investing the $27 million as shown in Table 14.

If they follow the plan of Table 14, the $250 million that can be invested in R&D will be reallocated as follows:

- Although less will be invested in research in the near term ($60 million versus $70 million), proportionally more will be invested in high-risk efforts (62% to CNS, versus 33%).
- The $20 million remaining in cardiovascular is already allocated to moderate-risk efforts; a total of $27 million of low-risk ("un-entrepreneurial") approaches has been eliminated.

Table 14. Potential Reallocation of Lucida R&D Budget

Research	New Total: $60 million
CNS	$37 million (+ $12 million)
CV	$20 million (– $10 million)
AI	$3 million (– $17 million)
Development	**New Total: $190 million**
CNS (Phase III)	$105 million (+ $10 million)
CV (Phase II)	$15 million (+ $5 million)
AI (Phase III)	$70 million

- The proportion of high-risk development will remain about the same (slightly more than half) in the near term.
- Overall, about 57% of R&D will be allocated to high-risk programs and projects in the near term. This is still less than the targeted two-thirds, although the proportion in research itself will be consistent with the desired archetype.

In any move, it is impossible to match the ideal proportions of the archetype immediately. At Lucida, it will take at least several years of careful reallocation, coupled with rigorous evaluation of the uncertainty and equivocality (hence, risk) of all proposed research programs, to reach the desired proportions.

Both the Board of Directors and the strategic leadership understand this. They also appreciate that the real test of people's commitment to the strategic vision is how "swiftly and smoothly" they can accommodate the severe disruptions of organizational life. For example, all of the money for near-term investment will come from a set of research programs in the cardiovascular and antiinfective areas that will have to be stopped. At least 115 people in R&D must be reassigned, and everyone will be affected during the transition. In addition, a number of new scientists with expertise in the selected future core technologies must be recruited and integrated.

To implement the change, the leadership involve everyone in R&D. From their discussions, a number of recommendations for the transition period are set forth. All R&D scientists will be offered a number of options:

- short sabbaticals at university and other laboratories, to develop new skills

- intense, in-house training with experts in the selected future core technologies
- part-time academic coursework in topics related to these technologies
- cross-training within R&D
- apprenticeship to newly hired scientists with important tacit knowledge in the selected future core technologies.

The decision to include all R&D staff, not just those immediately affected by the terminated programs, is made to alleviate a "them" versus "us" problem in the organization. In other words, even scientists working in efforts that will receive increased funding may be able to cross-train in another function, or to take a short leave to develop new skills, as long as the option does not disrupt the transition plan.

The Board, the strategic leadership, and the scientists are determined to maintain forward momentum on critical programs and development projects. In some cases, they decide staff should be grouped and housed separately, as if they are semiindependent contractors, until certain project milestones are reached (e.g., filing the New Drug Application). Wherever this is appropriate, a plan for reintegrating the scientists is developed. Thus, a number of scientists may move a project forward and then take a sabbatical to retrain, while others may stop their work immediately to gain new skills. Personal flexibility is key to allowing the work to continue and new recruits to be integrated with minimum disruption.

While the initial meetings are taking place within R&D, leaders of all infrastructure functions are asked to estimate how their operations will be affected by the change, and to submit plans for the respective operational support of the new archetype. Marketing is the most immediately affected by the decision to terminate antiinfective and cardiovascular research programs. Marketing management and staff meet often with their counterparts in R&D to discuss appropriate responses. (In fact, the $3 million that will remain in antiinfective programs is decided on the basis of those discussions.) In addition, marketing staff, like those in R&D, prepare a set of recommendations for their own retraining, to support the shift in research emphasis.

Less immediately affected by the new archetype, but with higher cost implications, is production. The new emphasis on gene technologies has major consequences for the types of plants required. The amount needed to upgrade existing facilities and build new ones is so substantial that it becomes clear that Lucida's profitability goals will not be met for the next several years. Discussions between the leadership and the Board result in a detailed plan showing the increase in internal investment and the expected revenues (and their timing) from the new research programs. R&D leaders realize that they will have to hold the total budget at $250 million *for at least 2 years,* given the obligatory expenses in marketing and production. The $27 million freed up by terminating cardiovascular and antiinfective research will be the only resource available, until the CNS therapies currently in Phase III begin to produce revenue.

From *Entrepreneur* to *Manager*

One of Solar Imaging's competitors is "Firm M"—*Master Medical Inc.*—a medical device company preparing to enter the big league. At the time of the competitive analyses in Chapter Five, Master Medical received about $1 billion in annual sales from both diagnostic devices and hospital supplies, and had invested about $80 million in diagnostic device R&D. The strategic leadership planned to increase R&D investment to that of the largest companies, around $350 million, by using proceeds from the sale of noncore (i.e., hospital supply) businesses.

The strategic decision to change competitive position is only being considered because device sales have been very profitable. Master Medical has been growing at about 15% per year for the past 5 years, principally from these products. The firm also has a sizable market share in several lines attractive to managed care organizations. By divesting other businesses and building up the device divisions, the leadership are convinced that Master Medical will be able to compete head on with the largest firms and succeed.

When the hospital supply business is sold, the Board agrees to invest $200 million in device R&D and the remainder in supporting functions, particularly manufacturing. With these resources, the leadership can now begin the process of changing competitive position by moving from the *technology entrepreneur* to *technology manager* archetype. (Their technology identity—to remain on the leading edge—will not change.)

At the start, science and technology resources in device R&D are invested as follows: 56% in internal capacity, which is consistent with the new archetype, and 61% in high-risk efforts (telemetry). The latter proportion is quite large compared with the modal firm (two-thirds in high- plus moderate-risk efforts). But, as I explained in Chapters Five and Six, the leadership have explicitly decided to invest heavily in the near term in telemetry, in order to gain competitive advantage (i.e., the third generic technology strategy investment choice).

In preparation for the change, Master Medical R&D staff have been evaluating a number of technologies that meet the three criteria for selected future core technologies and support the new archetype. Two clear choices are information technologies (moderate uncertainty/risk), including expert system software for clinical decision support, and fabrication technologies (low uncertainty/risk), that will assist their development engineers with prototype design in support of reduced manufacturing costs.

With these future technology options decided, the leadership at Master Medical will have to allocate the $200 million in new and expanded efforts in such a way that the proportions in R&D map those of the desired archetype. They develop a 4-year plan for allocation to the three risk categories as shown in Table 15. There will be temporary inconsistencies between the actual investment proportions within Master Medical and the desired proportions of the archetype during the transition. But, if implementation is successful, the resulting proportions will be: 30% in high-risk, 40% in moderate-risk, and 30% in low-risk efforts. Note that the future proportion allocated to high-risk R&D will be reduced considerably from the current 61%.

The strategic leadership in Master Medical face a different process for building the experience resource from the

Table 15. Master Medical Current and Planned Future R&D Budgets

Technology (Risk)	Current R&D: $80 million	Future R&D: $280 million
Telemetry (high)	$49 million	$84 million (+ $35 million)
Ultrasound (moderate)	$20 million	$62 million (+ $42 million)
Information technologies (moderate)	—	$50 million
Portable X-ray devices (low)	$11 million	$24 million (+ $13 million)
Fabrication (low)	—	$60 million

leadership in Lucida. In Lucida, the money for investing in new technologies and building the experience resource will come from the termination of existing activities. In Master Medical, the monetary resource will come from the sale of a division. Thus, in Master Medical the emphasis will be on recruiting new people rather than retraining and/or reassigning existing scientists, and morale is quite positive because of the expansion of R&D and prospects for continued success of the organization.

Although information and fabrication technologies are new to R&D, there is sufficient in-house expertise to evaluate and then integrate these capabilities. The leadership decide to acquire capacity (which will become in-house capacity), as well as to enter several strategic alliances, including one with a big-league competitor. The reason for the latter alliance is to monitor the future competition. The other alliances will be with firms outside the medical device sector.

Acquisition candidates will be sought primarily in the information technologies. The company has supported numerous academic groups at several universities and expects to be able to acquire at least two startups from these interactions immediately. At the end of 4 years, the planned investments in internal and external capacity in the three categories of risk will be as shown in Table 16.

The goal is to invest equally in internal and external capacity. Of the amount to be invested in-house ($140 million), more than 80% ($46 million + $70 million) will be allocated to high- plus moderate-risk efforts. Of the amount to be invested in external capacity ($140 million), about 40% will be allocated to low-risk efforts. The leadership will invest heavily in internal capacity in information technologies over the near term, to reduce the levels of uncertainty and equivocality within Master Medical faster than their competitors are able to reduce them. This is appropri-

Table 16. Master Medical Planned Future Investments in Internal and External Capacity

Risk (Technology)	Internal Capacity	External Capacity
High (telemetry, $84 million)	$46 million	$38 million
Moderate (IT, ultrasound, $112 million)	$70 million ($50 million IT)	$42 million
Low (fabrication, portable X-ray, $84 million)	$24 million	$60 million

ate to the near-term development of competitive advantage in a core technology, as discussed earlier. The proportion that will be needed to maintain their competitive position over the long term will be less (i.e., over the long term, the proportion will be consistent with the *technology manager* archetype).

A crucial issue facing the leadership is ensuring not only the technical but also the cultural fit of the acquisitions and of new recruits with Master Medical. They realize that people's values must support the desired archetype and technology identity (to remain on the leading edge). Thus, they are careful to assess candidates from that perspective as well. There is no guarantee they will judge correctly in all instances, but at least they have made this an explicit consideration.

In terms of business and R&D vector coordinate adjustment, again the process and issues at Master Medical are different from those at Lucida. Some of these differences arise from the much larger relative cost of manufacturing medical devices. To succeed in the big league, Master Medical will require state-of-the-art facilities. Most of the revenues from the current product lines will be devoted to building up a manufacturing capacity (in-house plus external) that will provide reduced throughput times, higher quality, and lower costs than at competing firms. Relatively less will need to be invested in marketing and other infrastructure functions, to support the enhanced products.

From *Pursuer* to *Manager*

The third illustration of an appropriate change in competitive position is provided by a hypothetical company not discussed in the earlier chapters. "Jensen-Brockmann" (referred to as "Jens-Bro") is a pharmaceuticals company formed nearly a decade earlier by a merger. At that time, Jensen had a large market share in gastritis therapies, and Brockmann had a strong presence in anesthetics.

Although the merged corporation has been profitable, a recent assessment of R&D by an outside consultant revealed a nearly dry research "pipeline." Most of the leadership's attention in the years following the merger was devoted to constructing new R&D facilities and reorganizing therapeutic area efforts. Less was paid to the caliber of research or to the focus of scientists' approaches.

At the time of this analysis, Jens-Bro sales have been flat (around $8 billion) for several years, and investment in R&D is $800 million. Several of their major products are over-the-counter gastritis medicines, under fierce competition and with diminishing margins. Moreover, several compounds will come off patent soon, drastically reducing future sales potential. The leadership are being pressured by shareholders to improve the company's lagging competitive position.

A portfolio review indicates that R&D resources within Jens-Bro are invested as shown in Table 17. Cancer and neurologic injury research takes place within a "new technologies" department. Including that department and all of the above areas, about 30 research programs are being

Table 17. Jens-Bro Current R&D Investments

Therapeutic Area	Research Budget	Development Budget	Total R&D
Cancer	$50 million	$20 million	$70 million
Neurologic injury	$30 million	—	$30 million
Diabetes	$80 million	—	$80 million
Atherosclerosis	$70 million	$50 million	$120 million
Anesthetics	$100 million	—	$100 million
Gastritis	$170 million	$150 million	$320 million
Antiinflammatory	$80 million	—	$80 million

supported. Currently, the only late-stage (Phase III) development project is in gastritis. The Phase III effort is almost complete, and approval of the drug is expected to follow soon after submission of the New Drug Application (the drug will be number five in its class when it is launched).

After reviewing this allocation, Jens-Bro leadership concede that they have, indeed, slipped to a pursuer position: 88% of overall resources are allocated to low- and moderate-risk efforts. Only cancer and neurologic injury efforts are considered high risk. Moreover, about three-quarters of the total allocated to low- (anesthetics, gastritis, antiinflammatory) and moderate-risk (diabetes, atherosclerosis) efforts is accounted for by licensed-in compounds.[3]

A series of painful and intense retreats with outside consultants and Jens-Bro R&D staff culminates in the decision to try to compete in the major league as a genuine *technology manager.* The firm has the potential scale but not the correct allocations of science and technology resources to do so. The leadership also agree that their technology identity—to remain on the leading edge—must be explicit.

Discussions between the Board and the senior leadership result in a window of 5 years during which the company will not be expected to meet certain business objectives, in order to allow the required shift of R&D investments to take place. After this period, performance standards will be reevaluated. All employees are notified of the decision, and the leadership ensure that there is frequent, open, and candid communication within R&D about the ensuing activities.

For the next 8 months, scientists are involved in an evaluation of every effort within R&D. Criteria for judging programs and projects include medical need, scientific caliber, market size, potential for the tightly cost-constrained health care environment, as well as fit with the desired archetype and technology identity. At the same time, a special task force of in-house plus external expert scientists assesses technologies that will meet the three future core technologies criteria and support the new archetype.

The outcome of these analyses is daunting. R&D programs and projects (including efforts on the in-licensed compounds) that should, on the above bases, be terminated affect 300 full-time equivalent (FTE) staff. The selected future core technologies, on the other hand, will require a minimum of 50 FTE new scientists with the latest knowledge, skills, and capabilities in molecular biology, growth factor production, information technologies, biosensors, and combinatorial chemistry. The task force recommends that a number of these new recruits be assigned to the former "new technologies" department, now designated "core technologies."

By the end of 5 years, the leadership plan to have the firm's science and technology resources invested as shown in Table 18. The recommended total invested in R&D will be $960 million, admittedly less than the major league firms will have invested 5 years from now. But the proportions allocated to the three risk categories will be more consistent with the managerial archetype: 39% to high-risk efforts (e.g., cancer, neurologic injury, biosensors); 28% to

Table 18. Jens-Bro Planned Future R&D Investments

Therapeutic Area	Research Budget	Development Budget	Total R&D
Cancer	$100 million	$50 million	$150 million
Neurologic injury	$90 million	$30 million	$120 million
Biosensors*	$100 million	—	$100 million
Atherosclerosis	$70 million	$50 million	$120 million
Addiction	$50 million	—	$50 million
Emphysema	$80 million	$30 million	$110 million
Anesthetics	$60 million	$50 million	$110 million
Gastritis	$50 million	$150 million	$200 million

*"Biosensors" is not a therapeutic area but a core technology that will be used across therapeutic areas.

moderate-risk efforts (e.g., atherosclerosis, addiction, emphysema); 33% to low-risk efforts (e.g., anesthetics and gastritis). In addition, they plan to reduce the proportion allocated to internal versus outside capacity from nearly two-thirds to slightly less than one-half.

However, during the next several years, the division of resources between research and development will be very different from that of other large firms in the pharmaceuticals sector (generally one-third in research and two-thirds in development). At Jens-Bro, the division will be nearly the opposite: 63% in research, 37% in development. The reason for this situation is twofold: there are few current compounds to be moved forward, and a heavy investment is needed to leverage the firm's research capabilities in core technologies. This is a precarious position, of course. The payback from research may be 10 years away, and the company might not survive that interim.

The leadership estimate that, 3 years after the desired reallocation shown in Table 18 is reached, another $300 million will be required for new compounds in development, with $300 million more being required in another 3 years. By the end of 10 years, assuming that discovery scientists are successful, the division of money between research and development activities will be more in line with the competition.

This is the plan, but it is still only paper. Building the experience resource at Jens-Bro into a competitive advantage is an extraordinarily difficult task. Because of the number of scientists who will be affected by the change, a human resources consultant specializing in R&D organizations is brought in. The goal is no layoffs, although this may be improbable.

In any event, because staff have been involved from the beginning, there are no surprises that can paralyze work. The human resources consultant, in concert with those in R&D, articulates the overriding objectives as: keep morale up, minimize disruption, move designated programs and projects forward as fast as possible, and retain those who are crucial to leverage Jens-Bro in the selected future core technologies. Some of the options they consider include:

- offer the "reskilling" alternatives provided at Lucida NeuroSciences
- offer early retirement incentives
- spin off separate businesses in which some staff could choose to work.

Each option has its downside, but at least there is agreement among employees that the company is acting in their best interest and is being very fair.

Another problem at Jens-Bro is the sheer size of R&D. In addition to dealing with the consequences of the archetype change, the leadership must also reorganize so that smaller and more informal work groups are the norm. One of the reasons that the company "slipped" to the pursuer position was the rigidity of the R&D structure. The leadership appreciate that the activities of the newly structured work groups must be consistent with the desired R&D vectors. They will have to continually balance the requirement for independent and lively team efforts with that of internal consistency.

CONCLUSION

Common Issues: Time To Implement, Perils of Implementing

There are two issues common to each change in competitive position: duration and (related to this) vulnerability. Even at Master Medical, with ample monetary resources and positive morale, there is an effective implementation period of about 6 years to analyze and select future core technologies and invest in these new efforts. Jensen-Brockmann illustrates perhaps the worst case; the plan for changing competitive position from *technology pursuer* covers more than a decade, *assuming all goes well*. Simply anticipating the move is likely to slow R&D work of its own accord, as scientists worry about their place in the new organization. Progress is never linear: programs and projects often fail, despite the best planning. Time frames of 4 or 5 years become time frames of 10 or more years.

Because of the length of time it takes to implement such a change, each firm is very vulnerable to the effects of turbulence as well as to competition during that period. There are also unavoidable inconsistencies that will still be perilous. For example, at Jens-Bro, the leadership must overinvest in research to the apparent detriment of development, because they are also reducing the amount allocated to outside capacity. If they licensed-in compounds during the interim, that would defeat the purpose of the move. Moreover, competition for good licensing possibilities would be so intense that the cost could be prohibitive.

At Lucida, the period of transition is also a time of fierce rivalry with two companies: Biotech 1 and Company D (Chapter Five). Some might recommend postponing the change—*but there is never a time without high penalties for failure.* That is why I emphasized that the leadership should ensure that their firm does not need to change competitive position. For most biomedical firms, this is an undertaking warranted only because long-term survival is not possible otherwise.

Differences

The case studies also illustrate the expected differences among archetype moves. Changing competitive position by moving from entrepreneur to manager, involving as it does a sizable addition to R&D, requires the most careful attention to cultural fit. Internal consistency, a key requisite to changing position successfully, will be jeopardized unless those who are evaluating new recruits assess not only their technical competence but also the values informing their technology identity. This is addressed in Chapter Eight in more detail.

Changing competitive position by moving from pursuer to manager requires the most careful attention to staff morale. Multiple programs and projects will have to be terminated, not because there is insufficient funding but because they are (in this strategic context) *not good enough.* The objectives set by the human resources consultant at Jens-Bro—keep morale up, minimize disruption, move designated programs and projects forward as fast as possible, and retain those who are crucial to leverage the firm in the selected future core technologies—are very difficult to achieve.

Finally, changing competitive position from quasi-manager to entrepreneur requires the most careful attention both to the caliber of science and to the risk-willingness of scientists. Defining the leading edge of relevant sciences and technologies involves continual experimentation at the leading edge, and frequent failure. There must be people in the organization who are "at" that edge in terms of their competence and connection to knowledge networks. In addition, as the example of Solar Imaging illustrated (Chapter Six), helping people to confront their own risk aversion is a crucial step in bringing about internal consistency with all that being a niche player demands.

NOTES

1. This is taken from the article cited in Chapter Six by J. A. Weisbach and W. H. Moos, "Diagnosing the Decline of Major Pharmaceutical Research Laboratories," *Drug Development Research 34* (March 1995), 243–259 (p. 250).
2. Also taken from Chapter Six is this reference to R. A. Goodman and M. W. Lawless, *Technology and Strategy,* Oxford University Press, New York, 1994.
3. The designation of levels of risk is based on the study cited in Chapter Five (A. M. Sapienza and I. Innami, "The Uncertainty of Pharmaceutical Research," presented at Technology Assessment and Forecasting Conference, Zurich, Switzerland (1993)).

CHAPTER EIGHT

ASSESSING THE FIT OF EXTERNAL KNOWLEDGE SOURCES

Because of rapidly advancing science frontiers, no biomedical firm can remain proficient in core technologies, over the long term, by itself. To achieve and sustain competitive advantage, merger (or acquisition), strategic alliances, and/or new recruits will be required. The purpose of this chapter is to provide criteria for assessing the fit of external knowledge sources and to illustrate the application of these criteria using example firms from prior chapters.

THE CONTEXT

Turbulence in the scientific and technologic dimension arises primarily because of rapidly advancing biomedical science frontiers. Radical discoveries such as those that have followed elucidation of the structure of DNA are providing enormous momentum to the field, increasing the rate of scientific and technologic change and the corresponding leadership challenge. "[As] the rate of change increases, the complexity of the problems that face us also increases."[1]

As fast as science and technology frontiers advance, so too must the capability of the biomedical firm. Despite the speed of change, effectively guided and supported knowledge-capturing activities within R&D can keep pace. Well-respected scholars of this subject have noted that even "a constant percentage accretion to the base [of knowledge] becomes exponential total growth. . . . [And] once a firm obtains a knowledge-based competitive edge, it becomes ever harder for competitors to catch up."[2]

To sustain a constant "accretion" to the experience resource in the context of rapidly advancing frontiers, however, the leadership must make use of knowledge sources external to the biomedical firm. This is not in contradiction to the warning against out-sourcing or sharing capacity in core technologies (Chapter Four). Rather, it is an explicit acknowledgment that, under conditions of turbulence, building and maintaining the experience resource *requires bringing external knowledge sources in-house.* New knowledge must be captured continually. To achieve and then sustain competitive advantage in the context of rapidly advancing science frontiers, the leadership must include sources such as other organizations (merger or strategic alliance) and new recruits.

It is possible to reduce the scope of new knowledge that must be captured, by strategic decisions that narrow the firm's domain. For example, the leadership could decide to exit a therapeutic area, thus decreasing the number of scientific disciplines in which scientists must be proficient. But, it is never possible to eliminate the need for new knowledge entirely. As I have emphasized in prior discussions, the *technology pursuer* is not a defensible archetype in the biomedical industry.

Which firms to acquire or partner with, and which talent to recruit, are extremely important strategic choices. To help in making these choices, I provide a set of four criteria in the following section. I do not mean to imply that these are the only bases for selection. Other considerations are important—geographic location, financial consequences, and so on—but they are beyond the scope of this discussion. I assume that marketing, manufacturing, and other infrastructure functions will require external knowledge sources, but that assessment is also beyond the scope of this discussion (although the same criteria would apply).

ASSESSMENT CRITERIA

Let me make clear that the criteria that follow apply only to supplemental capacity, to additions to core technologies. When I discuss merger or acquisition as a means of capturing knowledge, I refer to organizations that will constitute a minor portion of the firm's total science and technology resources. Merger or acquisition of equally sized firms requires assessment according to issues raised in Chapters Five and Six. I must also emphasize that assessment of current knowledge-capturing activities in core and noncore technologies should be based on the principle of internal consistency (Chapter Five) as well as the earlier discussion of technology (Chapter Four).

Below, I describe each criterion for judging fit (or not) as if it were discrete; in fact, they constitute an interdependent set of traits that selected knowledge sources must possess. If even one criterion is not met, competitive advantage will not be achieved. Knowledge may be captured, but not in a way that gives the biomedical firm a competitive edge.

Future Domain

Domain encompasses products or services, populations to whom the products or services will be provided, and technologies by which they will be produced. Over the 10- to 20-year timeframe of the strategic trajectories, there will be events and issues that will affect each aspect of the biomedical firm's domain. Safety, efficacy, or economic results will make some products or services more or less attractive. Changing epidemiology and demographics (populations) will make some products or services more or less important. Emerging technologies will increase the possible options within the technological matrix, and so on. These events and issues will be surfaced by ongoing environmental analyses and weak signal monitoring activities (Chapter Three). The fit of external knowledge sources must be based upon their ability to support the expected future domain.

The first criterion for judging fit is:

1. *Will the candidate knowledge source, in people's best judgment, support this biomedical firm's future domain?*

Competitive Uniqueness

In addition to supporting future domain, candidate knowledge sources should be relatively unknown to the industry; that is, they should not be obvious to the competition. If the set of candidates that the leadership are judging is on every other company's scouting list, those sources will not provide a competitive edge. Leaders who have a sophisticated understanding of technology (Chapter Four) are more likely to cast their net very widely and early, so the candidate sources can provide unique advantage.

The second criterion for judging fit is:

2. *Will the candidate knowledge source, in people's best judgment, provide a unique competitive advantage?*

Knowledge Sufficiency

Candidate knowledge sources will have been identified because it is presumed that they can add to the biomedical firm's capability. But, the funding available to be invested in them must be sufficient to build the experience resource in a meaningful way. Too little invested—even if it is consistent with the archetype—is, like too late invested, a waste of money. This is especially problematic if the leadership plan to invest in more than one external source at a particular time.

The third criterion for judging fit is:

3. *Will the intended investment in each candidate knowledge source, in people's best judgment, be sufficient to build the experience resource meaningfully?*

Values

Of course, it is not solely knowledge that is captured. Selected organizations and individuals also bring their own values. The principle of internal consistency requires that the values inherent in the external knowledge sources be consistent with the values that underlie the biomedical firm's technology identity.

The dilemma facing everyone involved in this assessment is that one cannot ask the direct question: "What are your values?" As a well-respected anthropologist stated in so many words, people cannot be asked what they value but must be observed valuing.[3] That is, the assessment of fit must be based upon indirect or oblique measures. Take the case of judging new recruits, assuming that they have passed the former screens. Is there evidence that they have experimented and failed? Is there evidence that they have learned from failure? Is there evidence that their career has been "eccentric" to the norm (e.g., did they ever study or work in very different disciplines), and so on?

The questions relevant to an organization are similar. Based on actual performance data and the observations of outside experts, has the work of the organization really been on the leading edge? Has it defined the leading edge? What is the evidence that the organization encouraged experimentation and supported failure? What is the evidence that learning occurred? Has the organization's evolution been eccentric to the norm?

A useful "litmus test" of values is for in-house scientists to observe the behavior of potential recruits and people in organizations being considered for acquisition. What interests them? What excites them? What kinds of questions do they ask about the scientists' own organization?

These observations provide useful insights into the values actually held. For example, taking (intelligent) chances reflects (1) a risk-willing attitude, (2) the value of risk willingness, and (3) an orientation towards science and technology of at least remaining on, if not defining, the leading edge. Lack of failure, of an individual or of an organization, should warn people that safety (and mediocrity) may be prized.

The fourth and final criterion for judging fit is:

4. *In people's best judgment, is there evidence that the values actually held by the candidate knowledge source are consistent with this firm's desire to remain on or to define the leading edge?*

Figure 37 summarizes the criteria in a flow chart. The answer, "No," to *any* question requires the search for candidate knowledge sources to resume. All criteria must be

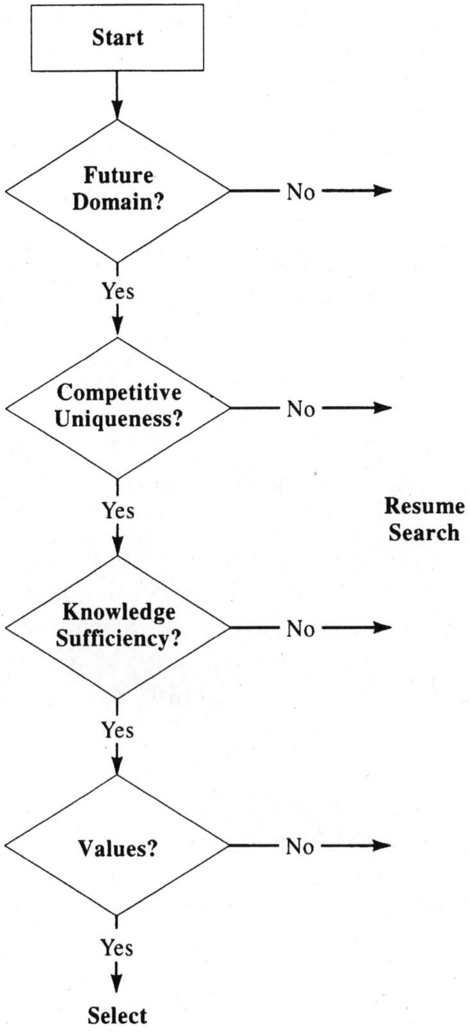

Figure 37. Criteria for assessing fit.

met by candidates that are selected, if competitive advantage is to be achieved.

APPLYING THE CRITERIA

Using example firms from prior chapters, I illustrate the application of these criteria to the three major sources of external knowledge that may be brought in-house:

- *Master Medical*—acquisition of information technology startups
- *Lucida NeuroSciences*—research strategic alliances
- *Jensen-Brockmann*—new recruits.

As with other illustrations, they are not intended to do justice to the time and intensity of effort that are required nor are they meant to cover the entire spectrum of issues that must be considered. They do suggest a systematic approach to assessing and selecting external knowledge sources.

Master Medical: Acquiring Startups

One of the strategic choices facing the leadership as they move Master Medical from the *technology entrepreneur* to the *technology manager* archetype is how to build information technology (IT) capability. As I described in the prior chapter, IT plus expert systems have been selected as one set of future core technologies that will support Master Medical's change in competitive position.

The leadership have been funding a number of academic groups at technical universities, and several startup organizations have emerged to date. Master Medical had negotiated right-of-first-refusal contracts, to acquire any or all of the organizations once they had legally separated from the university. In Chapter Seven, I noted that the leadership expected to be able to acquire two of these startups, for a total of $50 million (Table 16), immediately.

The initial pool of candidate knowledge sources consists of five organizations that the leadership assess according to the criteria presented above.

Future Domain. Analyses of the environment relevant to the medical device sector globally underscore the strategic importance of being able to provide diagnostic information not only on individual patients but also on panels of patients. Master Medical scientists have designed a prototype telemetry system that can collect diagnostic data from multiple devices and remote locations, evaluate them on the basis of given protocols, and prepare an array of reports for clinical and financial users.

From their ongoing experience with the candidate organizations, R&D scientists believe that all five startups can support at least part of this aspect of Master Medical's expected future domain. Two of the organizations have designed infrared communications devices for local linkup; three of them have designed applicable expert systems. Thus, one choice (based on hardware/software compatibility) is to select one startup from each category. (Remote-site linkup will make use of existing local- and wide-area network providers.)

Competitive Uniqueness. Master Medical's right-of-first-refusal contracts have provided effective hedges against competitors' acquiring these startups. However, on further analysis, it becomes clearer to the leadership that not every organization will provide unique advantage. There are a number of startups with relatively similar infrared devices being assessed by other medical device firms in the big league. The software startups, on the other hand, appear to be relatively unknown in this sector.

On that basis, the leadership decide to continue with the assessment of the software organizations but to cast their net more widely with regard to equipment acquisitions. Because of competitive pressures, they agree to a deadline of 3 months for the search. During this time, in-house scientists will try to determine if software alone can provide a unique competitive advantage to Master Medical's future products.

Knowledge Sufficiency. Given three possible acquisitions, the leadership must judge what capability can be brought in-house for a particular amount of money. If they reserve $20 million for a potential equipment startup (a reasonable amount, given what competitors are willing to pay), what software capability might they gain for $30 million?

Each software company is offered for about $25 million. One startup consists of 30 full-time equivalent (FTE) staff; another consists of 28 FTEs. Both organizations are focused on expert systems relevant to Master Medical's prototype devices. The third organization consists of 12 FTEs with similar expert systems capacity plus 20 FTEs working on other software advances that may be relevant to future product lines.

During discussions about these options, Master Medical scientists argue there are benefits to all candidates. The first two would, they believe, sizably increase Master Medical's capacity in expert systems. The third, although not as large in that one arena, might add value to more aspects of R&D.

In support of the latter option (i.e., acquiring the third startup), scientists point out that 12 FTEs is a considerable number to be devoted to their particular effort. What is less clear to everyone is how much knowledge the remaining staff would add to current and future projects. At the end of the day, people agree that Master Medical's IT capability will be increased meaningfully by any one of these three startups.

Values. For Master Medical staff, this is the most important aspect of fit. Scientists in R&D are very confident of their own ability to compete face-to-face with companies in the big league. They are insistent that people in these startups not only provide new technical capacity but also evince the desire to remain on or to define the leading edge of their sciences and technologies.

Several days are set aside for scientists from each software startup to reside on the Master Medical campus and meet formally and informally with prospective colleagues. Although the company supported the groups for some years, this is the first time that scientists at Master Medical will be able to interact, observe, and consider the orientation towards science and technology that might be brought with the acquisition.

When the visits are over, Master Medical scientists meet again to share their impressions. Interestingly, there is ready agreement that, although scientists from the first two organizations are technically capable, they appear to be very cautious and less risk-willing than believed to be appropriate to Master Medical's upcoming competitive challenge. People from the third startup appear to be more inclined to take chances. They readily discussed attempts that had failed as wonderful opportunities to learn. They were more upbeat about competing with the big league; in fact, they were eager to do so.

The outcome of these discussions is the recommendation that the leadership invest in the latter organization.

Master Medical scientists acknowledge that there may be too few people from the startup assigned to other efforts (i.e., not the expert systems software), to provide a meaningful increase in the experience resource. But the 12 FTE scientists will provide unique advantage and a meaningful increase in IT capacity. Most importantly, all of the people from that startup appear to share Master Medical scientists' values with regard to risk taking and experimentation, as well as to enjoying large challenges. On that basis alone, people agree this acquisition is attractive.

Of Note

During the assessment process, it became evident to the leadership that their "net" had not been cast early enough, or wide enough, to identify hardware startups that would provide Master Medical with a unique competitive advantage. If one of the original two hardware organizations were acquired, knowledge would be captured, but not in a way that gave Master Medical an edge on the big-league firms. Fortunately, the software startups might make up for this lack.

Failure of a candidate knowledge source to meet the *competitive uniqueness* criterion arises for three reasons, any or all of which may apply.

First, the leadership may not understand the properties associated with technology well enough to judge weak signals accurately (Chapter Four). They will thus invest at a late-growth stage of the life cycle rather than at an emerging or early growth stage. In other words, Master Medical staff may have invested too late in the academic groups. Second, the leadership may understand technology and the implications of the life cycle but be too risk-averse to invest at the appropriate stage, implying internal inconsistency between actual values and desired technology identity (Chapter Six). Third, the leadership may understand technology, be risk-willing, but be poorly served by the strategic intelligence and weak-signal-monitoring activities in their organization (Chapter Three). What appears to be the universe of candidates is in fact only a partial set because of limited data.

It would be important for the future success of any biomedical firm for staff to debrief the assessment process and consider what lessons should be learned.

Lucida NeuroSciences: Research Strategic Alliance

As described in Chapter Seven, the leadership at Lucida NeuroSciences are involved in intense efforts to realign their science and technology investments so that they are consistent with the *technology entrepreneur* archetype. One of their decisions is to increase the allocation to CNS research by $12 million (Table 14), almost one-third again as large, specifically in emerging-stage technologies.

Although all R&D scientists are involved in knowledge-capturing activities, such as in-house training and sabbaticals, the leadership realize that the time required to build their experience resource in this therapeutic niche is too long for the competitive situation (cf. Chapter Five). In discussions with the scientists, they decide to invest $10 million of the $12 million in CNS in a research strategic alliance. Their objective is to capture new knowledge as quickly as possible, to regain ground lost by their earlier inconsistency of investment.

A group of in-house scientists is charged with identifying a pool of alliance candidates that will then be formally assessed according to the four criteria. After consultation with outside academic researchers and other experts in the field, the task force proposes four companies—one biotechnology company (not their current competitor from Chapter Five) and three pharmaceuticals companies.

Future Domain. Broader definitions of the biopharmaceutical product to include more than a compound (i.e., the pill) are a crucial part of the future landscape of health care, according to analyses of the environment. Lucida scientists believe that the information and education aspects of medicines will be at least as important for buyers (patients, physicians, reimbursers) of the twenty-first century. Given aging of populations and large cohorts of people with chronic conditions, "disease management" is a service that Lucida staff believe is crucial for them to provide. The four companies identified as alliance candidates have developed reputations for innovation in relevant areas.

Competitive Uniqueness. Most firms in the industry are involved in similar discussions about redefining the product. However, according to their strategic intelligence, Lucida scientists contend that no firm yet has mounted serious efforts in CNS disease management. There are pilot activities in other therapeutic areas, like diabetes and asthma, that are proving successful in attracting managed care organizations. If they are able to enter an appropriate alliance soon, the leadership believe they can put a considerable competitive distance between Lucida and other entrepreneurial firms. Based on this criterion, all four companies are considered appropriate.

Knowledge Sufficiency. Lucida is a small company in comparison with its alliance partners, and the leadership realize that $10 million is a very small amount to invest. Will it make a meaningful difference to their CNS experience resource?

The candidate knowledge sources offer capabilities that are very different (although related to Lucida's future domain) and, consequently, difficult to compare in order to answer that question. All four companies are conducting research in areas important to Lucida's CNS efforts, but each brings additional capacity that makes it attractive as an external knowledge source. For example, the biotechnology firm offers highly respected pharmacoeconomics techniques. These will be useful for Lucida scientists when their compound enters preclinical development. One pharmaceuticals company has a long-standing arrangement with a pharmaceutical benefits management firm that provides unique information on disease management. Such information may offer an advantage to Lucida scientists during their discovery efforts. Another owns a clinical facility and offers unusually detailed insight into future treatment regimens. The third pharmaceuticals company has a global reputation for their gene-signaling techniques.

For the amount that Lucida can invest in the alliance, each firm appears able to bring a modest but meaningful and distinct addition to their knowledge base. Scientists are initially divided in their opinion as to which alliance might leverage them most quickly. After a number of intense discussions and reevaluation of proposed alliance terms, they decide to eliminate the third pharmaceuticals company from further consideration at this time. They believe that $2 million invested in gene-signaling training for in-house scientists will provide benefit equivalent to $10 million invested in an alliance with this firm for its gene-signaling expertise.

Values. Lucida NeuroSciences formerly exhibited the perilous inconsistency between vision and technology identity. Despite choosing an explicit niche strategy (vision), the leadership had continued to invest as if the firm were a *technology manager,* with approximately equal amounts in the three risk categories. Of insufficient scale to compete with the big league and with too little allocated to high-risk efforts, their effective technology identity was to remain on, rather than to define, the leading edge. People spoke like entrepreneurs but acted in a more risk-averse manner.

The effort required at Solar Imaging to align vision and values (Chapter Six) was also required at Lucida. Perhaps because of that experience, Lucida staff are apprehensive about their ability to discern the values actually held by scientists in the candidate firms. Many of Lucida's scientists acknowledge their continuing unease about how risk-willing a niche company must be. However, with the help of a consultant, that unease is put to good use.

Lucida scientists are advised that what they should be aware of is how *eccentric* the potential partner appears in comparison with themselves. Rather than try to discern similarity with values they themselves are struggling to internalize, they are counseled to focus on observing differences. For example, scientists from the second pharmaceuticals company (with the clinical facility) are perceived to be totally willing to gamble on efforts to "push the envelope" in their CNS research. On the other hand, scientists from the biotechnology company (with pharmacoeconomics expertise) appear more similar to Lucida staff in terms of risk aversion. They show little interest in broadening the definition of their product to include much more than the compound plus minimal educational materials. Scientists from the first pharmaceuticals company (with access to a pharmacy benefits management firm) are also observed to be more similar to Lucida staff.

The discussions on values prove useful to the continuing development of Lucida scientists, in terms of providing deeper insight into their own struggles for internal consistency. They agree that the firm that feels most eccentric will benefit Lucida enormously, by what one scientist calls its "positive momentum" (the pharmaceuticals company with the clinical facility). Using the same metaphor, scientists decide that candidate organizations similar to Lucida at the present time are, by that token, likely to exert negative inertial forces on the desired technology trajectory. On this basis, the leadership decide to enter an alliance with the pharmaceuticals company that owns the clinical facility.

Of Note

In their assessment of fit with Master Medical values, R&D scientists sought confirmation that candidate knowledge sources were *similar* to themselves. In the case of Lucida, that fit was better determined by *difference.* The reason for the apparent paradox is that Master Medical staff are already characterized by the technology identity consistent with their desired competitive position. On the other hand, Lucida staff are still in transition (from the identity consistent with the *technology manager* to the identity consistent with the *technology entrepreneur*). They are correct to use the alliance not only to build the CNS experience resource but also to provide a positive force for their own desired technology trajectory.

Too often, people assess fit by how closely the other organization (or person) resembles their own (themselves). In certain situations, similarity is the right metric. In most others, as in Lucida's, the right metric is the perceived eccentricity of the organization or individual to the norm.

Jensen-Brockmann: New Recruits

Perhaps the most difficult strategic shift to accomplish successfully is the move from *technology pursuer* archetype.

Jensen-Brockmann, as described in Chapter Seven, illustrates the enormity of this undertaking. The plan for reallocating science and technology resources covers more than a decade, assuming all goes well. Moreover, the leadership must initially overinvest in high-risk efforts, compared with the modal *technology manager,* simply to build an experience resource that will allow Jens-Bro to be viable in that archetype. In other words, they must allocate a disproportional amount to highly uncertain research in-house, at the same time that they struggle with all the other issues involved in changing competitive position.

Analyses of R&D revealed that a minimum of 50 FTE new scientists were needed, to ensure knowledge "accretion" in molecular biology, growth factor production, information technologies, biosensors, and combinatorial chemistry. Jens-Bro leadership realize they have a formidable task. They must hire the right recruits and integrate them as fast and as smoothly as possible into an organization that is unavoidably in flux. In hiring people, they must try to maintain the appropriate balance between (1) scientists with industry experience but perhaps less current theoretical knowledge, and (2) more recent graduates with cutting edge knowledge and technical skills but no industry experience. (The right balance will differ from firm to firm.) Based on discussions with their scientists, the leadership decide to use the assessment criteria to design a special job description, with components suggested by each criterion. They do not expect that every candidate will meet every component, but selected recruits must meet at least one or two for each criterion. This job description will supplement the company's own formal documentation.

Future Domain. Analyses of the future health care environment underscore for the leadership that Jens-Bro products must demonstrate clear cost advantages to managed care buyers. Moreover, there must be a steady source of such products for the company to succeed in the face of ongoing compression of product life cycles. In the parallel job description, people decide that new recruits should therefore be assessed in terms of their:

- understanding of at least the fundamentals of pharmacoeconomics
- demonstrated ability to contribute to speed-to-market project management techniques
- familiarity with the evolution of population-based reimbursement methodologies
- environmental-scanning facility, as evidenced by current use of advanced electronic search engines.

Competitive Uniqueness. Given the expected future domain, most companies in the pharmaceuticals sector are seeking recruits with the same discipline-based capacity that Jens-Bro leadership are seeking. If they can recruit scientists with the above additional competencies, that will provide the firm with some competitive advantage. But more is needed. To gain a competitive edge, the following are added to the job description:

- doctoral studies in respected but unlikely universities (e.g., better known for their humanities education but with one or two strong science programs; public institutions with one or two strong science programs; institutions from nations not currently well represented in particular disciplines, and so on)
- work experience in organizations outside, but relevant to, the health care industry
- if currently employed in academic institutions, holding a rank lower than associate professor

- if holding a rank of associate or full professor, with recent publications on still-contested issues.

Knowledge Sufficiency. Scientists at Jens-Bro have come to appreciate that the knowledge that will serve them best, competitively, is *tacit knowledge.* They decide that particular backgrounds of recruits will be more likely than others to add meaningfully to the tacit aspect of Jens-Bro's experience resource, such as:

- formal or informal apprenticeships in novel techniques, as opposed to formal study only of such techniques
- independent work using skills that Jens-Bro needs, rather than working only in a subordinate position to people with those skills
- training in radically different techniques
- substantive experience in functions related to drug discovery (e.g., time spent in project management, or drug development, or marketing, and so on).

Values. Because values must be observed, the remaining components of the supplemental job description for Jens-Bro new recruits are contained in interview questions. In addition to having scientists actually spend time with recruits who pass the first three screens, certain questions are believed to provide useful insight into values actually held. The interview questions include:

- "Tell me about the best job you ever held." (What was most challenging to them? What was most exciting? Why did they love it?)
- "What was the worst job you ever held?" (What did

they dislike? Why did they dislike it? What did they do about it?)

- "What are the good points and bad points about what you're doing now?" (What does this add to their answers to the first two questions?)
- "Tell me about an effort of yours that did *not* succeed." (Have they experimented? Have they learned from failure? What have they learned? Are they discouraged by failure?)
- "What do you believe is your best work?" (Do they

Future Domain

- Fundamentals of pharmacoeconomics
- Speed-to-market project management techniques
- Population-based reimbursement methodologies
- Advanced electronic search engines

Competitive Uniqueness

- Unlikely university doctorates
- Work outside health care industry
- Instructor or assistant professor
- Publishing in still-contested areas

Knowledge Sufficiency

- Apprenticeships in novel techniques
- Independent work
- Radically different techniques
- Experience in other related functions

Values

- What is the best job you ever had?
- What was the worst job?
- What effort did not succeed?
- What is your best work?

Figure 38. Supplemental job description—Jens-Bro.

enjoy "pushing the envelope"? What has challenged them? Is there evidence of single-minded persistence?)

When the job description and interview questions are completed (Figure 38), Jens-Bro scientists are coached in interview skills. They role-play situations and discuss what they learn, until they are confident about their ability to pick (as far as is ever possible) the right candidates. Most of them are very surprised by the depth of understanding that is required to solicit important information, especially candidates' actually held values and orientation towards science and technology. They agree that, although there are no guarantees, they are in a much better position to select the right knowledge source than they would ever be without such preparation.

Of Note

Assessment of candidate organizations, like Master Medical's startups and Lucida's research alliance partners, also entails assessment of the individuals involved. However, performance of the organization is taken as a proxy for the caliber of its staff. When individuals are being recruited, there may be no corresponding organization whose performance can be evaluated as well. In fact, good individuals often come from poorly performing firms (because they are frustrated in their own attempts to succeed).

Jens-Bro scientists are wise to come up with the supplemental job description that, in addition to highlighting fit or not with the assessment criteria, also highlights what I have termed the *human aspects* of an individual's personal competencies.[4] People are usually able to discern the technical aspects that an individual brings (education, training, etc.)

and match them to the technical aspects of the job (required techniques and skills). But few can, without training and coaching (as in Jens-Bro), identify the human aspects such as orientation (attitudes) towards science and technology.

Of course, hiring the right people is just the beginning. As I said earlier, the leadership at Jens-Bro still have to integrate the recruits quickly and smoothly into an organization in transition. This requires a commitment of time and informal as well as formal training efforts.

NOTES

1. This statement was made almost two decades earlier by R. Ackoff in *Creating the Corporate Future*, Wiley, New York, 1981, p. 4.
2. This is an important statement for those who must ensure that the biomedical firm keeps pace with rapidly advancing science frontiers. J. B. Quinn, P. Anderson, and S. Finkelstein , "Leveraging Intellect," *Academy of Management Executive 10*, No. 3, (1996), 7–27. The citation is from p. 8.
3. This is based on the work of B. Malinowski, *Argonauts of the Western Pacific*, EP Dutton and Co. Inc., New York, 1953. What Malinowski actually said (p. 20) is the following: "[There] is a series of phenomena of great importance which cannot possibly be recorded by questioning or computing documents, but must be observed in their full actuality. Let us call them the *imponderablia of actual life*."
4. R. E. Boyatzis ("Competence at Work," in A. J. Stewart (Ed.), *Motivation and Society*, Jossey-Bass, San Francisco, CA, 1982) described a "motivated" organization as one in which there is a good fit among (1) personal competencies, (2) job demands, and (3) organizational characteristics. I took this model and divided each of the three spheres into technical and human aspects, based on my observation that most of us are much more facile at matching technical personal competencies with technical job demands (A. Sapienza, *Managing Scientists*, Wiley, New York, 1995).

CHAPTER NINE

COMPETITIVE TECHNOLOGY STRATEGY: SOME IMPLICATIONS FOR TEAMS

The purpose of this chapter is to address some of the implications of the new model of strategy for program and project management, and for program and project teams. The discussion is divided into two parts. In the first, general implications from each of the prior chapters are presented. In the second, implications specific to technology transfer are described. It should be clear by the end of the chapter that the strategic process of determining, setting, and continuously adjusting the coordinates of the R&D vector becomes actionable at the program and project level. Program and project teams implement the biomedical firm's competitive technology strategy.

"STRATEGY IS EMBEDDED IN PROJECTS"

Early on, I described R&D programs and projects as "booster rockets." They are both the sources of energy that propel the biomedical firm towards its strategic vision and the means by which R&D vector coordinates are adjusted. As I described in Chapter One, every program and project can be represented by a vector. The sum of the discovery efforts and research programs defines the research vector, the sum of the development projects defines the development vector, and the sum of both defines the R&D vector. Over time, the R&D vectors describe a series of points that define the firm's technology trajectory.

The title of this introductory section, "strategy is embedded in projects," is meant to emphasize the strategic importance of program and project management. The author of that statement explained:[1]

> There are two ways in which strategy is embedded in projects. . . . The pre-project phase encompasses the activities by which the firm establishes its priorities and identifies the technologies it expects to be involved in future projects. . . . The post-project phase governs the ability of the organization to learn from one project to the next, to assess the effectiveness of organizational or technical approaches employed, and to make improvements that, over a series of projects, will leverage project experience into a more innovative system.

I also made the point in Chapter One that program, project, R&D, and corporate decisions are inextricably linked. Just as there are no "vision-neutral" values, so, too, there are no "vector-neutral" decisions. The technology trajectory can, certainly, be knocked off course by events in the external environment (cf. solar flares, meteorites). But, it can also be *sent* off course by poor R&D decisions. These result in programs and projects that are in inappropriate areas, or are underfunded, or are badly managed so they fail to achieve their objectives. Any program or project that is poorly executed affects the R&D vector, which affects the technology trajectory.

Biomedical programs and projects always face the prospect of scientific and technical wrong turns and apparent dead ends. That is the nature of science as people explore the interface between the known and the unknown. Failure is unavoidable and an opportunity for learning. Effective R&D leadership take this into account, so that failure is not catastrophic for the firm. Ineffective R&D leadership, on the other hand, produce inappropriately deployed rockets, underfueled rockets, and badly aimed rockets. Poor decisions at the program and project level can propel the organization as far from the strategic vision as if it were struck by a meteorite.

The objective of this chapter is to discuss the implications of the new model of strategic trajectories that I believe are most important for program and project teams. In the first part, I present some general implications from each of the prior chapters. In the second part, I present some implications specific to technology transfer. My intent is to make clear that the strategic process of determining, setting, and continuously adjusting the coordinates of the R&D vectors becomes actionable at the program and project level. Program and project teams implement the biomedical firm's competitive technology strategy.

SOME GENERAL IMPLICATIONS

Strategy: *Communication*

Good strategic decisions, those that result in strategic success, depend upon good communication, no matter how large or how small the organization (cf. the feedback loops depicted in Figures 5 and 6). Because of the tight linking of program, project, R&D, and corporate decisions, team members must continually seek out and interpret data and communicate relevant information wherever needed: to other members, team leaders, R&D directors, people in other divisions and functions, and corporate management. Similarly, information that affects the teams must be communicated in timely fashion from wherever it originates. Because of the tight linking, decisions at all levels must be informed decisions.

At a minimum, good communication means that:

- People are well-informed by means of sophisticated information technologies and data sources that are accessible on a "want to know" basis[2]
- They are candid in their discussions
- They freely challenge and surface assumptions
- They exercise good judgment
- They discuss the business of the biomedical firm, without regard for function, title, status, etc., and without fear of reprisal.

One of the reasons that I described strategic leadership as a "most important and terrifying" task is that program and project decisions have strategic consequences. And the effectiveness of these decisions depends on good organizational communication. Poor decisions resulting from poor communication propel the biomedical firm away from its strategic vision. Program and project team members are, because the organization has been turned inside-out in this new model of strategy, especially reliant on and responsible for good communication and effective decisions. They work, so to speak, on the skin of the rocket. In that position, they need timely information to respond appropriately, and they are likely to be first to perceive certain data crucial to the firm (discussed further below). Team leaders must be "fluent" in the languages of both science and business, as well.

Team members are also one means by which the firm can act instrumentally. They are uniquely able, by virtue of their position on the skin of the rocket, to shape a part of the environment. Instrumental program and project decisions, based on good communication, ensure that the biomedical firm influences, shapes, and creates its future.

Environment: *Instrumental Intelligence*

If program and project teams work on the skin of the rocket, they are likely to be first to recognize something in space of strategic importance to the firm, particularly in (but not limited to) the scientific/technologic and epidemiologic dimensions. I said in Chapter Three that everyone in the organization must be responsible for gathering strategic intelligence, but R&D teams are likely to be first to find out about other organizations' scientific success or failure, emerging problems in clinical trials, and so on.

Because of this, teams must operate at the highest level of strategic sophistication. The function of R&D teams is not simply to gather strategic intelligence but to appreciate how the information may be used instrumentally. Hence, if team members are not strategically sophisticated, it does not matter if R&D and corporate management are.

It is also strategically important that team members be actively involved in the respective "invisible colleges" that, taken together, constitute the biomedical firm's membership in the knowledge network. Such membership can be a critical barrier to the entry of competing firms, and it can provide useful information regarding external knowledge sources (see below).

With regard to the environment, R&D teams provide crucial input to the question: "What is the impact of (an event or issue that has been identified)?" (cf. Figure 18). Team members' involvement in (1) identifying key components of dimensions, (2) assessing levels of turbulence, (3) determining impact, (4) gathering intelligence, and (5) monitoring weak signals can make the difference between intelligence and *instrumental intelligence.* Only instrumental intelligence helps the biomedical firm to influence, shape, and create its future.

Technology: *Person-Embodied*

Tacit knowledge is "person embodied," and core technologies are by definition those in which there is a sizable amount of tacit knowledge, in which their "software" ("performance programs stored in individual memory") is more important than their "hardware."[3] Every team member should be viewed as a repository of potentially powerful competitive advantage to the biomedical firm. If a team member from a core technology leaves, the knowledge content of that technology is reduced and the organization's competitive advantage weakened to a greater or lesser degree. (I am not advocating lifetime employment but effective R&D leadership.)

A core technology, in other words, is nothing more (and nothing less) than a group of people whose valuable talents must be kept and developed in-house as much as possible. Outsourcing core technologies reduces the amount that could be invested in educating, training, and building teams to the biomedical firm's competitive advantage.

By the same token, the levels of uncertainty, equivocality, and risk associated with a technology are social constructs. A technology within R&D has a high level of uncertainty because scientists on the team do not yet fully understand the raw materials and/or the processes by which they are transformed into some output. A technology within R&D has a high level of equivocality because scientists on the team do not yet fully agree on the technology's use, role, and scope. It is within teams—either the firm's or the competitors'— that understanding is accumulated (uncertainty is reduced) and agreement is reached (equivocality is reduced). Effective program and project leadership is a necessary condition for speedy reduction of uncertainty and/or equivocality, to the biomedical firm's competitive advantage.

As I discussed in Chapter Four, there are different tactics by which the two independent properties are reduced. When the level of uncertainty is high, program and project leaders must work intensively to support wide-ranging search (i.e., information collection and processing) activities of team members. These activities can be undertaken by team members independently of one another. But, when the level of equivocality is high, team leaders must bring the members together, in face-to-face (transactive) communication, until agreement is reached. This will be addressed again in the section on technology transfer.

Determining Competitive Position: *Quantifying the Experience Resource*

Vector metrics used to determine the competitive position of the biomedical firm are one means of quantifying the experience resource. Tangible science and technology resources, especially expert personnel, constitute potential competitive advantage. Turning these resources into real advantage is accomplished by investing in team members' knowledge, skills, and capabilities. Teams constitute the firm's experience resource, because experience is "person embodied."

The knowledge, skills, and capabilities of program and project team members are partly quantified by the *origin, magnitude,* and *direction* of each program and project vector. Investing in teams (*origin* and *magnitude*) and setting them to work on programs and projects with particular risk levels (*direction*) help the biomedical firm to learn, adapt, change, and renew over time.[4] This is what was meant by the ability of an organization "to learn from one project to the next, to assess the effectiveness of organizational or technical approaches employed, and to make improvements that . . . will leverage project experience into a more innovative system."[5] The experience resource is both person embodied (in each team member) and organization embodied (in the innovative system that is the R&D organization). This resource becomes a competitive advantage when the firm's teams are, in salient respects, superior to competitors' teams. Keeping this resource superior, of course, involves replenishing and supplementing the team by means of new and more technologically advanced recruits (Chapter Eight). In the dynamic situation of biomedical R&D (cf. Chapters Three and Four), effectively integrating, training, and mentoring new recruits are crucial to the firm's success.

Determining the competitive position of R&D by means of vector metrics is an attempt to assess the quantity of the experience resource. All measures of quantity are imperfect, because information in a turbulent environment is imperfect. But they are necessary, nonetheless. Another crucial function of team members, on the rocket skin, is to help make imperfect information on competitors' experience resources more meaningful as strategic intelligence. If team members are aware of the caliber of scientists in competing firms and the approaches to problems they are taking, their thoughtful judgments will inform and enlighten the leadership's shared understanding of what the vector metrics *mean* in terms of the firm's competitive position.

Consistency and Competitiveness: *Changing Values*

Person-embodied attitudes towards science and technology must be in accord, over the long term, with the technology identity defined by the strategic vision. When there is inconsistency between vision and identity, change is neither quick nor easy. Given that attitudes reflect what people value, when there is inconsistency at the program and project level, people on the teams must change what they value.

At Lucida NeuroSciences, for instance, there was inconsistency between the strategic vision (to be a niche company, or *technology entrepreneur*) and actual technology identity (essentially, to remain on the leading edge). Scientists on program and project teams had to be made aware of the inconsistency, and appreciate the consequences for the technology trajectory, before the science and technology investments and the experience resource could be shifted appropriately.

Paradoxically, the more effective the leadership are at ensuring consistency of vision and technology identity (or attitudes and archetype), the more difficult the situation when temporary *in*consistency is required. When it is appropriate to pursue the leading edge of a technology *pro tem* (because of a strategic change in domain, for example), R&D scientists with an orientation of defining or remaining on the leading edge will find it difficult to reduce their efforts.

Conceptually, the decision to let capability in "old" technologies lag that in "new" technologies is straightforward.

At a human level, the level of the affected teams, this decision should be antithetical to what scientists value. But, in order for the biomedical firm's investment to be well spent, less effort will be required than they are used to making. Scientists must effectively change their values for a time.

If the firm's technology identity is to remain on the leading edge, but resources must be allocated to reduce uncertainty or equivocality faster within the firm's teams than within competitors' (i.e., to define the leading edge of future core technologies), the human problem is similar. Scientists on the team will be asked for a qualitatively much greater effort, to gain understanding of raw materials and conversion processes (or to reach agreement on use, role, and scope of the technology) more rapidly than competitors. Again, scientists must effectively change their values for a time.

In both of these situations, when temporary inconsistency is required, program and project leaders should try to move scientists off the activity as soon as feasible.

Changing Competitive Position: *Reassigning People*

Of all the efforts described in this book, changing competitive position has perhaps the biggest impact on program and project teams. Investing, and shifting investment in, technologies is conceptually straightforward. Investment must be reduced immediately in "old domain" technologies that are in the mature stage. For "old domain" core technologies in earlier stages of the life cycle, investment must be only as much as will help the biomedical firm to maintain competitive position. (This amount usually represents no or reduced growth in program and project funding.) For selected future core technologies, on the other hand, investment must be such that levels of uncertainty and/or equivocality are reduced faster within the firm's teams than they are within competitors' teams.

Although the steps are conceptually straightforward, these technologies are embodied in people. Changing competitive position is achieved in large measure by reassigning program and project team members. Consider the example of Lucida NeuroSciences. Of the $20 million supporting antiinfective research programs, only $3 million are to be preserved. Of the $30 million invested in cardiovascular research programs, $10 million support "old domain" efforts. Shifting investment in technologies, to move Lucida from the *quasi-manager* to *technology entrepreneur* archetype, entails reassigning 115 people within R&D.

The situation is even more dramatic in Jensen-Brockmann, which the leadership are attempting to move from the *technology pursuer* to *technology manager* archetype. Portfolio decisions will affect 300 people (Chapter Seven). In addition, the precarious position of the company has resulted in the Board's decision to allow R&D a grace period of 5 years in which to demonstrate that Jens-Bro can compete. This is a doubly negative situation for scientists: they face program/project termination, reassignment, and a limited timeframe for building up new knowledge, skills, and capabilities. At both Lucida and Jens-Bro, scientists must change their values. Moreover, many of these scientists probably have years of experience in a particular field. Unless thoughtfully managed, being reassigned to another field could be damaging to their careers.

The impact of changing competitive position on program and project teams is part of the reason why I do not recommend this undertaking, unless it is absolutely necessary for the biomedical firm's survival.

Assessing External Knowledge Sources: *Finding and Integrating People*

Competitive technology strategy depends, as was pointed out in the prior chapter, on continually capturing new knowledge that will sustain competitive advantage. "Capturing" knowledge, of course, means finding the right people from external sources and integrating them successfully in program and project teams. Whether the strategic choice is to recruit individuals, to merge with or acquire another organization, or to enter a strategic alliance, so-called knowledge accretion in R&D entails finding and integrating people.

All four criteria for judging the fit (or not) of external knowledge sources are concerned with finding and integrating people. The first criterion (support of the firm's *future domain*) can be restated as two questions: Do the people have knowledge, skills, experience, and capabilities that are germane to the firm's future products and markets? Is there any evidence that they are able and willing to share their knowledge with the program and project teams in which they will be integrated?

The second criterion (*competitive uniqueness*) can be restated: Are the people familiar to many other firms or a few? The more closely linked team members are to their respective invisible colleges, the more readily they will spot people who bring a unique competitive advantage, because they are relatively unfamiliar to competitors. Membership in the knowledge network is not only a potential barrier to entry but also a vehicle for discovering unique external knowledge sources.

With regard to the third criterion (*knowledge sufficiency*), external knowledge sources have been identified because they are presumed to be capable of supplementing the firm's current capability. However, the sufficiency criterion is determined only partly by the source's scientific reputation, vitae, and stated accomplishments. More important are the judgments of team members about the demonstrable added value they believe the people will provide. At the end of the day, knowledge sufficiency depends on new recruits' willingness to share their knowledge and experience with program and project team members (similar to the first criterion, above).

Finally, in terms of the fourth criterion (*values*), external sources bring not only knowledge (person embodied) but also values (person embodied). As I described in the prior chapter, the litmus tests for value "fit" are the initial interactions between external people and team members. Internal scientists must observe people's behavior, to discern what interests and excites them. From these observations, they will make some inferences about what values people may actually hold. Team members must then decide if those values are consistent with their own desire to remain on or to define the leading edge.

SOME IMPLICATIONS FOR TECHNOLOGY TRANSFER

In every biomedical firm, the work of producing an output consists of a series of operations that can be described by discrete steps and decision criteria (e.g., go/no go). Customarily, after the initial idea-generating process, a number of teams are formed (often a separate team per step) to bring the idea to market as a tangible good and/or intangible service.

The time required to go from idea to product ranges from a few to many years, depending on the industry sector—fewer years for some medical devices and many years for some biopharmaceutical medicines. However, although the sector and type of product may dictate the general time parameters, there is much variability among firms. Development times may be much longer for some firms within the same sector than for others.[6]

Such variability is an important factor in the economic success or failure of the company. Products that are not among the first two or three to market may not generate enough revenues for the company to cover their costs. Speed-to-market is critical in the biomedical industry. The firm that can bring products to market more quickly than other organizations in the relevant sector has a distinct competitive advantage. From the perspective of this chapter, program and project teams must ensure that the time from idea development to product introduction is as short as possible. This is no small accomplishment, as the following discussion shows.

Each program and project, each step in the idea-to-product process, has its own life cycle.[7] Figure 39 illustrates a very simplified operation consisting of six major steps, two of which are conducted in parallel (2 and 3) and the remainder sequentially. The purpose of this figure is to highlight the inherent discontinuities—the interfaces between steps—involved in bringing a product to market (there will be $N - 1$ interfaces or discontinuities, where N is the number of discrete, sequential, program and project steps). Each interface involves "technology transfer," defined as "the iterative movement of . . . applied knowledge via one or more communication channels. . . ."[8] Each interface is also a potential "crack" through which critical knowledge, skills, and experience can "fall." If that happens, speed-to-market, a competitive advantage, is lost.

In other high-technology industries, observers have commented that effective management of the interfaces—the technology transfer, or hand-offs from one program and project team to another—is at least as important as ef-

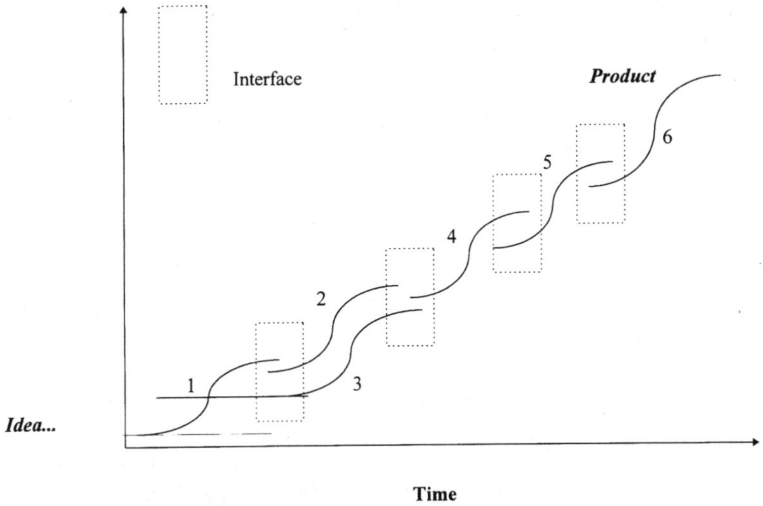

Figure 39. From idea to product.

fective management of the project itself. Managing hand-offs well, in fact, has been described as "key to rapid product and service development."[9] Because of the high knowledge content of biomedical technologies, managing technology transfer well is essentially managing communication of that knowledge well.[10]

If the same people were involved from idea to product, if the operation were a seamless process within one site, technology transfer would not be so difficult. But complex biomedical technologies require many different disciplines (i.e., different people are involved in each team); a number of distinct, sometimes regulation-required steps; and often more than one geographic location. Because each technology transfer, each hand-off from one project step to the next, can be a rate-limiting step, I have designed a set of communication strategies that minimize time and maximize effectiveness. These strategies are based on the likely levels of uncertainty and equivocality that will be experienced by the follow-on team. Those levels, in turn, can be inferred from two attributes of that follow-on team: (1) unfamiliarity (how unfamiliar are people on the follow-on team with the effort?), and (2) heterogeneity (how diverse in terms of disciplines represented is the follow-on team?).

As I described in Chapter Four, uncertainty means that something remains unknown. Thus, level of unfamiliarity with the effort should be a good predictor of level of uncertainty. Similarly, level of discipline diversity (i.e., heterogeneity) should be a good predictor of level of equivocality. The greater the discipline differences within the team, the less likely members are to have reached agreement on the use, role, and scope of the applied knowledge that (perhaps along with a tangible product) is being handed over to them. Again, uncertainty and equivocality are social constructs; they are attributes of the scientists on the teams.

Given three levels (low, moderate, and high) of unfamiliarity and heterogeneity for the sake of practical simplicity, there are nine appropriate communication strategies illustrated in Figure 40. Managing technology transfer well (quickly and effectively) entails understanding the two attributes of the follow-on team and selecting the appropriate communication strategies for the hand-off. For example, in cell 1 of the figure (low unfamiliarity, low heterogeneity), only the standard dossier should be necessary for the follow-on team. In cell 2, however, the moderate heterogeneity of the follow-on team suggests that both the standard dossier and one or a few meetings between the teams will be required. Because of the moderate equivocality, bringing people together in transactive (two-way) communication is necessary. (Equivocality is only reduced by bringing people together until agreement is reached.)

As the level of unfamiliarity (uncertainty) increases, so,

Team Heterogeneity (Discipline Differences)			
Team Unfamiliarity With Effort	**Low**	**Moderate**	**High**
Low	(1) Dossier only	(2) Dossier plus one or a few team meetings	(3) Dossier plus many team meetings
Moderate	(4) Dossier plus additional documentation	(5) Dossier, additional documentation, one or a few meetings	(6) Dossier, additional documentation, many team meetings
High	(7) Dossier plus sizable additional documentation	(8) Dossier, sizable documentation, one or a few team meetings	(9) Dossier, sizable documentation, many team meetings

Figure 40. Appropriate communication stategies for speedy technology transfer.

too, does the documentation requirement. Moderate unfamiliarity (cells 4 through 6) cannot be reduced by the same dossier that was sufficient for cells 1 through 3; additional documentation is necessary. High unfamiliarity requires sizable additional documentation. Similarly, as the level of heterogeneity (and associated equivocality) increases, so, too, does the number of meetings. More meetings are required to reduce high equivocality than to reduce moderate equivocality.

Staying abreast of rapidly advancing science frontiers, and remaining on or defining the leading edge of relevant sciences and technologies, means that teams will be dealing with moderate to high uncertainty and equivocality, by definition. In terms of the figure, program and project teams will be operating predominantly in the lower quadrant (cells 4 through 9). For speedy and effective technology transfer, it is crucial that scientists understand the communication strategies contingent on those properties or attributes of the follow-on teams. Ensuring competitive advantage by the shortest possible time from idea to market depends on managing communication well.

One additional consideration is important. If facilities are in different geographic locations, managing the meetings required to reduce equivocality will be problematic. When it is economically unfeasible to bring entire teams together during the hand-off, the "sending" team has three options.[11]

First, several scientists from the follow-on team can be integrated into the prior team (generally during the mature phase of the earlier life cycle) and then sent back to their facility with the tangible product as well as the more crucial tacit knowledge and experience of those team members. Second, scientists who have worked in the follow-on function (e.g., preclinical development) can work with the prior team and then take the product plus knowledge and experience to the next team and remain with them until their unfamiliarity and equivocality is reduced. Third, experienced scientists can be used in a consulting capacity, working with the prior and follow-on teams in the same way as just described.

From a pragmatic perspective, if the teams are geographically close, the first option may be the most sensible. As geographic distance increases, rather than remove scientists from their team for too long, it may be better to use the second or third options. However, if the latter are selected, scientists must go to the follow-on team as "missionaries" for the effort, in order to transfer "knowledge [and obtain] . . . buy-in" of those team members.[12]

NOTES

1. The quotation, as well as the title of the section, is taken from P. Adler, "Technology Strategy: A Guide to the Literatures," unpublished report, Department of Engineering and Engineering Management, Stanford University (1988), p. 47.
2. The normative statement that information should be provided on a "want to know" rather than a "need to know" basis comes from S. Jorgensen, I. Jensen, and M. Edwards, "Surviving as a Niche Player in the Pharmaceutical Industry," *Drug Development Research 30* (November 1993), 112–120.
3. These references are to W. Faulkner and J. Senker, *Knowledge Frontiers*, Oxford University Press, Oxford, UK, 1995 ("person embodied") and to D. Gerwin, "Relationships Between Structure and Technology," in P. S. Nystrom and W. Starbuck (Eds), *Handbook of Organizational Design*, Oxford University Press, Oxford, UK, 1981 ("performance programs"). Both sources were first cited in Chapter Four.
4. Also referred to in Chapter Four is D. J. Teece, G. Pisano, and A. Shuen, "Dynamic Capabilities and Strategic Management," Working Paper #90-8, Center for Research in Management, University of California, Berkeley (1990).
5. This is taken from Adler (Note 1, above), p. 47.
6. A benchmarking study by the Centre for Medicines Research (UK) of pharmaceuticals firms illustrated the variability in development times among firms, even in similar therapeutic areas. See *CMR News*, summer 1996, for an overview of the pilot benchmarking study of efficiency in drug discovery and development.
7. For more information on program and project life cycles (as well as other aspects of general project management), the reader is referred to *A Guide to the Project Management Body of Knowledge*, PMI Standards Committee, Project Management Institute, Upper Darby, Pennsylvania, 1996.
8. The quotation is from the same authors who studied technology transfer at the U.S. Microelectronics and Computer Technology Consortium (F. Williams and D. V. Gibson, "Introduction," in F. Williams and D. V. Gibson (Eds.), *Technology Transfer: A Communication Perspective*, Sage, Newbury Park, CA, 1990.
9. The title of P. B. Scott-Morgan's article, in fact, is "Managing Interfaces," *Prism*, second quarter (1991), 55–69. The quotation is from p. 63.

10. The importance of good communication for effective technology transfer is drawn from Williams and Gibson (Note 8, above).
11. The three options are drawn from D. Leonard-Barton, "The Intraorganizational Environment," in F. Williams and D. V. Gibson (Eds.), *Technology Transfer: A Communication Perspective,* Sage, Newbury Park, CA, 1990.
12. The need for scientists (in this case) to be "missionaries" is from D. Leonard-Barton (Note 11, above), p. 51.

INDEX